Water Power Engineering

Water Power Engineering

Naresh Lamba

RANDOM PUBLICATIONS
NEW DELHI (INDIA)

Water Power Engineering

ISBN 978-93-5111-427-7

Published in 2014 in India by

RANDOM PUBLICATIONS

Reprint 2018

4376-A/4B, Gali Murari Lal, Ansari Road
New Delhi-110 002
Phone : +91-11-43580356, +91-11-23289044
e-mail: randomexports@gmail.com, sales@randompublications.com, info@randompublications.com

Type Setting by: Friends Media, Delhi-110089
Printed at : Replika Press Pvt. Ltd.

Preface

The water of the oceans and water bodies on land are evaporated by the energy of the sun's heat and gets transported as clouds to different parts of the earth. The clouds travelling over land and falling as rain on earth produces flows in the rivers which returns back to the sea. The water of rivers and streams, while flowing down from places of higher elevations to those with lower elevations, loose their potential energy and gain kinetic energy. The energy is quite high in many rivers which have caused them to etch their own path on the earth's surface through millions of years of continuous erosion. In almost every river, the energy still continues to deepen the channels and migrate by cutting the banks, though the extent of morphological changes vary from river to river. Much of the energy of a river's flowing water gets dissipated due to friction encountered with its banks or through loss of energy through internal turbulence. Nevertheless, the energy of water always gets replenished by the solar energy which is responsible for the eternal circulation of the Hydrologic Cycle.

Hydropower engineering tries to tap this vast amount of energy available in the flowing water on the earth's surface and convert that to electricity. There is another form of water energy that is used for hydropower development: the variation of the ocean water with time due to the moon's pull, which is termed as the tide. Hence, hydropower engineering deals with mostly two forms of energy and suggest methods for converting the energy of water into electric energy. In nature, a flowing stream of water dissipates throughout the length of the watercourse and is of little use for power generation. To make the flowing water do work usefully for some purpose like power generation (it has been used to drive water wheels to grind grains at many hilly regions for years), it is necessary to create a head at a point of the stream and to convey the water through the head to the turbines which will transform the energy of the water into mechanical energy to be further converted to electrical energy by generators. Electricity from water is usually referred to as Hydro-Power, where the term 'hydro' is the Greek word for water and hydropower is the energy contained in water. It can be converted in the form of electricity through hydroelectric power plants. All that is required is a continuous inflow of water and a difference of height between the water level of the upstream intake of the power plant and its downstream outlet. Some nations have enough hydropower to become exporters of electricity. Switzerland, for example, exports electricity to neighbouring France and Italy. Nepal, Bhutan, Peru and Laos are similarly blessed with abundant hydro resources. Within India, Meghalaya is probably the only state generating hydropower more than its requirements and exports power to the neighbouring state of Assam.

The primary aim of this book is to present a succinct account of the essential features of this subject. It is being prepared by keeping in view the requirements of the students and academic professionals.

I thank all members of my team who have helped in the preparation of the book. My special thanks go to "Random Publications" who have published the book.

—Naresh Lamba

Contents

1

Introduction

The Civil Engineering Body of Knowledge is currently set forth in a proposal by the American Society of Civil Engineers (ASCE) entitled *Civil Engineering Body of Knowledge for the 21st century*. This proposal seeks to identify and implement improvements to the education and licensure process for civil engineers in the United States of America.

Current Status

In the United States, the body of knowledge necessary to obtain a license to practice engineering is defined by the laws or regulations of each state or territory. Most states currently have a standard that is a four step process. First, an individual must obtain a Bachelor's degree from a university program that is accredited by the Accreditation Board for Engineering and Technology.

A two-step examination process administered by the National Council of Examiners for Engineering and Surveying must be completed. The first eight-hour test is the Fundamentals of Engineering exam; the second, also eight hours long, is the Principles and Practice of Engineering exam. The other step is to work an apprenticeship, usually of four years in length, under an already-licensed engineer. The second exam is generally the fourth and final step; the fundamentals exam can be taken before or after the apprenticeship in most states.

The first state to regulate the practice of engineering was Wyoming in 1907. After that, ASCE established a model law for licensure. The last state to pass licensure laws for engineers was Montana.

Many states now require continuing education to maintain a license to practice engineering. In 1979, Iowa became the first. Since then about half of the states have added continuing education to their engineering laws.

History

The ASCE board of directors adopted a policy in 1998 (Policy Statement 465) that supported a change to make the master's degree be the first professional degree to enable practice of civil engineering. This proposed change was not widely accepted within the civil engineering profession and the policy was first revised in 2001 to support a requirement for a "master's degree or equivalent". It was revised again in 2004 to support "the attainment of a body of knowledge for entry into the practice of civil engineering at the professional level"

The ASCE board created a standing committee, the Committee on Academic Prerequisites for Professional Practice (CAP^3), charged with the implementation of Policy Statement 465. CAP^3 determined that the best implementation of PS 465 was to define the body of knowledge (BOK) that would form the foundation of the licensure process. CAP^3 in turn established the Body of Knowledge Committee which wrote the first (2004) and second (current, 2008) versions of the BOK.

Content the BOK

The body of knowledge defines twenty-four outcomes that make up the knowledge, skills and attitudes necessary to practice civil engineering. The outcomes are divided into three categories: foundational, technical, and professional.

- Foundational
  - Mathematics
  - Natural sciences
  - Humanities
  - Social sciences
- Technical
  - Materials science
  - Mechanics
  - Experiments
  - Problem recognition and solving
  - Design

  * Sustainability
  * Contemporary issues & historical perspectives
  * Risk and uncertainty
  * Project management
  * Breadth in civil engineering areas
  * Technical specialisation
- Professional
  * . Communication
  * Public policy
  * Business and public administration
  * Globalisation
  * Leadership
  * Teamwork
  * Attitudes
  * Lifelong learning
  * Professional and ethical responsibility.

The body of knowledge uses Bloom's Taxonomy to outline the necessary level of achievement for each of the twenty-four outcomes.

Implementation Status

ASCE has formed the BOK Educational Fulfillment Committee (BOKEdFC) to focus on the changes needed to engineering education. This committee is composed of representatives from universities with four-year civil engineering programs. NCEES considered the implementation of the BOK at their 2008 annual meeting and decided to establish a task force. The task force is provide an analysis of "(1) the potential educational, professional, regulatory, and economic impact of the master's or equivalent; and (2) any alternative solutions besides the master's or equivalent that could potentially address the challenge of better preparing engineering licensure candidates to enter the profession."

In 2008, Nebraska became the first to consider legislation requiring college-level education beyond the Bachelors degree as a requirement for a professional engineering license.

The legislation was not enacted, in part to testimony from engineering associations. The Nebraska section of the American Council of Engineering Companies stated that the new requirement might have made it more difficult for companies located in their state to hire and keep entry-level engineers.

Engineering Ethics

Engineering ethics is the field of applied ethics and system of moral principles that apply to the practice of engineering. The field examines and sets the obligations by engineers to society, to their clients, and to the profession. As a scholarly discipline, it is closely related to subjects such as the philosophy of science, the philosophy of engineering, and the ethics of technology.

The 19th Century and Growing Concern

Figure: *The first Tay Bridge collapsed in 1879. At least sixty were killed.*

As engineering rose as a distinct profession during the 19th century, engineers saw themselves as either independent professional practitioners or technical employees of large enterprises. There was considerable tension between the two sides as large industrial employers fought to maintain control of their employees. In the United States growing professionalism gave rise to the development of four founding engineering societies: ASCE (1851), the American Institute of Electrical Engineers (AIEE) (1884), ASME (1880), and the American Institute of Mining Engineers (AIME) (1871). ASCE and AIEE were more closely identified with the engineer as learned professional, where ASME, to an extent, and AIME almost entirely, identified with the view that the engineer is a technical employee. Even so, at that time ethics was viewed as a personal rather than a broad professional concern.

Turning of the 20th Century and Turning Point

When the 19th century drew to a close and the 20th century began, there had been series of significant structural failures, including some spectacular bridge failures, notably the Ashtabula River Railroad Disaster (1876), Tay Bridge Disaster (1879), and the Quebec Bridge collapse (1907). These had a profound effect on engineers and forced the profession to confront shortcomings in technical and construction practice, as well as ethical standards.

One response was the development of formal codes of ethics by three of the four founding engineering societies. AIEE adopted theirs in 1912. ASCE and ASME did so in 1914. AIME did not adopt a code of ethics in its history. Concerns for professional practice and protecting the public highlighted by these bridge failures, as well as the Boston molasses disaster (1919), provided impetus for another movement that had been underway for some time: to require formal credentials (Professional licensure in the US.) as a requirement to practice. This involves meeting some combination of educational, experience, and testing requirements.

Over the following decades most American states and Canadian provinces either required engineers to be licensed, or passed special legislation reserving title rights to organisation of professional engineers. The Canadian model is to require all persons working in fields of engineering that posed a risk to life, health, property, the public welfare and the environment to be licensed, and all provinces required licensing by the 1950s.

The US model has generally been only to require those practicing independently (i.e. consulting engineers) to be licensed, while engineers working in industry, education, and sometimes government need not be licensed. This has perpetuated the split between professional engineers and those in industry. Professional societies have adopted generally uniform codes of ethics.

On the other hand technical societies have generally not adopted these, but instead sometimes offer ethics education and resources to members similar to those of the professional societies. This is not uniform, and the question of who is to be held in the highest regard: the public or the employer, is still an open one in industry, and sometimes in professional practice.

Recent Developments

Efforts to promote ethical practice continue. In addition to the professional societies and chartering organisations efforts with their members, the Canadian Iron Ring and American Order of the Engineer trace their roots to the 1907 Quebec Bridge collapse. Both require members to swear an oath to uphold ethical practice and wear a symbolic ring as a reminder.

In the United States, the National Society of Professional Engineers released in 1946 its Canons of Ethics for Engineers and Rules of Professional Conduct, which evolved to the current Code of Ethics, adopted in 1964. These requests ultimately led to the creation of the Board of Ethical Review in 1954. Ethics cases rarely have easy answers, but the BER's nearly 500 advisory opinions have helped bring clarity to the ethical issues engineers face daily.

Currently, bribery and political corruption is being addressed very directly by several professional societies and business groups around the world. However, new issues have arisen, such as offshoring, sustainable development, and environmental protection, that the profession is having to consider and address.

General Principles

> *"Engineers, in the fulfillment of their professional duties, shall hold paramount the safety, health, and welfare of the public."*
>
> *—National Society of Professional Engineers*
>
> *"A practitioner shall, regard the practitioner's duty to public welfare as paramount."*
>
> *—Professional Engineers Ontario*

Codes of engineering ethics identify a specific precedence with respect to the engineer's consideration for the public, clients, employers, and the profession.

Many engineering professional societies have prepared codes of ethics. Some go back to the early decades of the twentieth century. These have been incorporated to a greater or lesser degree into the regulatory laws of several jurisdictions. While these statements of general principles served as a guide, engineers still require sound judgment to interpret of how the code would apply to specific circumstances. The general principals of the codes of ethics are largely similar across the various engineering societies and chartering authorities of the world, which further extend the code and publishes specific guidance. The following is an example from the American Society of Civil Engineers:

1. Engineers shall hold paramount the safety, health and welfare of the public and shall strive to comply with the principles of sustainable development in the performance of their professional duties.

2. Engineers shall perform services only in areas of their competence.
3. Engineers shall issue public statements only in an objective and truthful manner.
4. Engineers shall act in professional matters for each employer or client as faithful agents or trustees, and shall avoid conflicts of interest.
5. Engineers shall build their professional reputation on the merit of their services and shall not compete unfairly with others.
6. Engineers shall act in such a manner as to uphold and enhance the honour, integrity, and dignity of the engineering profession and shall act with zero-tolerance for bribery, fraud, and corruption.
7. Engineers shall continue their professional development throughout their careers, and shall provide opportunities for the professional development of those engineers under their supervision.

Obligation to Society

The paramount value recognised by engineers is the safety and welfare of the public. As demonstrated by the following selected excerpts, this is the case for professional engineering organisations in nearly every jurisdiction and engineering discipline:

- Institute of Electrical and Electronics Engineers: "We, the members of the IEEE, ... do hereby commit ourselves to the highest ethical and professional conduct and agree: 1. to accept responsibility in making decisions consistent with the safety, health and welfare of the public, and to disclose promptly factors that might endanger the public or the environment;"
- Institution of Civil Engineers: "Members of the ICE should always be aware of their overriding responsibility to the public good. A member's obligations to the client can never override this, and members of the ICE should not enter undertakings which compromise this responsibility. The 'public good' encompasses care and respect for the environment, and for humanity's cultural, historical and archaeological heritage, as well as the primary responsibility members have to protect the health and well being of present and future generations."
- Professional Engineers Ontario: "A practitioner shall, regard the practitioner's duty to public welfare as paramount."

- National Society of Professional Engineers: "Engineers, in the fulfillment of their professional duties, shall: Hold paramount the safety, health, and welfare of the public."
- American Society of Mechanical Engineers: "Engineers shall hold paramount the safety, health and welfare of the public in the performance of their professional duties."
- Institute of Industrial Engineers: "Engineers uphold and advance the integrity, honour and dignity of the engineering profession by: 2. Being honest and impartial, and serving with fidelity the public, their employers and clients."
- American Institute of Chemical Engineers: "To achieve these goals, members shall hold paramount the safety, health and welfare of the public and protect the environment in performance of their professional duties."
- American Nuclear Society: "ANS members uphold and advance the integrity and honour of their professions by using their knowledge and skill for the enhancement of human welfare and the environment; being honest and impartial; serving with fidelity the public, their employers, and their clients; and striving to continuously improve the competence and prestige of their various professions."

Whistleblowing

A basic ethical dilemma is that an engineer has the duty to report to the appropriate authority a possible risk to others from a client or employer failing to follow the engineer's directions. According to first principles, this duty overrides the duty to a client and/or employer. An engineer may be disciplined, or have their license revoked, even if the failure to report such a danger does not result in the loss of life or health.

Figure: *The Space Shuttle Challenger disaster is used as a case study of whistleblowing and organizational behavior including groupthink.*

In many cases, this duty can be discharged by advising the client of the consequences in a forthright matter, and ensuring the client takes the engineer's advice. However, the engineer must ensure that the remedial steps are taken and, if they are not, the situation must be reported to the appropriate authority. In very rare cases, where even a governmental authority may not take appropriate action, the engineer can only discharge the duty by making the situation public. As a result, whistleblowing by professional engineers is not an unusual event, and courts have often sided with engineers in such cases, overruling duties to employers and confidentiality considerations that otherwise would have prevented the engineer from speaking out.

Conduct

There are several other ethical issues that engineers may face. Some have to do with technical practice, but many others have to do with broader considerations of business conduct. These include:

- Relationships with clients, consultants, competitors, and contractors
- Ensuring legal compliance by clients, client's contractors, and others
- Conflict of interest
- Bribery and kickbacks, which also may include:
  - o Gifts, meals, services, and entertainment
    - a. Treatment of confidential or proprietary information
    - b. Consideration of the employer's assets
    - c. Outside employment/activities (Moonlighting).

Some engineering societies are addressing environmental protection as a stand-alone question of ethics.

The field of business ethics often overlaps and informs ethical decision making for engineers.

Case Studies and Key Individuals

Petroski notes that most engineering failures are much more involved than simple technical mis-calculations and involve the failure of the design process or management culture. However, not all engineering failures involve ethical issues. The infamous collapse of the first Tacoma Narrows Bridge, and the losses of the Mars Polar Lander and Mars Climate Orbiter were technical and design process failures.

These episodes of engineering failure include ethical as well as technical issues.

- Space Shuttle Columbia disaster (2003)
- Space Shuttle Challenger disaster (1986)
- Therac-25 accidents (1985 to 1987)
- Chernobyl disaster (1986)
- Bhopal disaster (1984)
- Kansas City Hyatt Regency walkway collapse (1981)
- Love Canal (1980), Lois Gibbs
- Three Mile Island accident (1979)
- Citigroup Centre (1978), William LeMessurier
- Ford Pinto safety problems (1970s)
- Minamata disease (1908–1973)
- Chevrolet Corvair safety problems (1960s), Ralph Nader, and *Unsafe at Any Speed*
- Boston molasses disaster (1919)
- Quebec Bridge collapse (1907), Theodore Cooper
- Johnstown Flood (1889), South Fork Fishing and Hunting Club
- Tay Bridge Disaster (1879), Thomas Bouch, William Henry Barlow, and William Yolland
- Ashtabula River Railroad Disaster (1876), Amasa Stone.

Earthquake Engineering

Earthquake engineering is the scientific field concerned with protecting society, the natural and the man-made environment from earthquakes by limiting the seismic risk to socio-economically acceptable levels. Traditionally, it has been narrowly defined as the study of the behaviour of structures and geo-structures subject to seismic loading, thus considered as a subset of both structural and geotechnical engineering. However, the tremendous costs experienced in recent earthquakes have led to an expansion of its scope to encompass disciplines from the wider field of civil engineering and from the social sciences, especially sociology, political sciences, economics and finance.

The main objectives of earthquake engineering are:

- Foresee the potential consequences of strong earthquakes on urban areas and civil infrastructure.

- Design, construct and maintain structures to perform at earthquake exposure up to the expectations and in compliance with building codes.

A properly engineered structure does not necessarily have to be extremely strong or expensive. It has to be properly designed to withstand the seismic effects while sustaining an acceptable level of damage.

Seismic Loading

Seismic loading means application of an earthquake-generated excitation to a structure (or geo-structure). It happens at contact surfaces of a structure either with the ground, or with adjacent structures, or with gravity waves from tsunami.

Seismic Performance

Earthquake or seismic performance defines a structure's ability to sustain its due functions, such as its safety and serviceability, *at* and *after* a particular earthquake exposure. A structure is, normally, considered *safe* if it does not endanger the lives and well-being of those in or around it by partially or completely collapsing. A structure may be considered *serviceable* if it is able to fulfill its operational functions for which it was designed.

Basic concepts of the earthquake engineering, implemented in the major building codes, assume that a building should survive a rare, very severe earthquake by sustaining significant damage but without globally collapsing. On the other hand, it should remain operational for more frequent, but less severe seismic events.

Seismic Performance Assessment

Engineers need to know the quantified level of the actual or anticipated seismic performance associated with the direct damage to an individual building subject to a specified ground shaking. Such an assessment may be performed either experimentally or analytically.

Experimental Assessment

Experimental evaluations are expensive tests that are typically done by placing a (scaled) model of the structure on a shake-table that simulates the earth shaking and observing its behaviour. Such kinds of experiments were first performed more than a century ago. Still only recently has it become possible to perform 1:1 scale testing on full structures.

Due to the costly nature of such tests, they tend to be used mainly for understanding the seismic behaviour of structures, validating models and verifying analysis methods. Thus, once properly validated, computational models and numerical procedures tend to carry the major burden for the seismic performance assessment of structures.

Analytical/Numerical Assessment

Seismic performance assessment or, simply, seismic structural analysis is a powerful tool of earthquake engineering which utilises detailed modelling of the structure together with methods of structural analysis to gain a better understanding of seismic performance of building and non-building structures. The technique as a formal concept is a relatively recent development. In general, seismic structural analysis is based on the methods of structural dynamics. For decades, the most prominent instrument of seismic analysis has been the earthquake response spectrum method which, also, contributed to the proposed building code's concept of today.

However, such methods are good only for linear elastic systems, being largely unable to model the structural behaviour when damage (i.e., non-linearity) appears. Numerical *step-by-step integration proved* to be a more effective method of analysis for multi-degree-of-freedom structural systems with significant non-linearity under a transient process of ground motion excitation.

Basically, numerical analysis is conducted in order to evaluate the seismic performance of buildings. Performance evaluations are generally carried out by using nonlinear static pushover analysis or nonlinear time-history analysis. In such analyses, it is essential to achieve accurate nonlinear modelling of structural components such as beams, columns, beam-column joints, shear walls etc. Thus, experimental results play an important role in determining the modelling parameters of individual components, especially those that are subject to significant nonlinear deformations. The individual components are then assembled to create a full nonlinear model of the structure. Thus created models are analysed to evaluate the performance of buildings.

The capabilities of the structural analysis software are a major consideration in the above process as they restrict the possible component models, the analysis methods available and, most importantly, the numerical robustness. The latter becomes a major consideration for structures that venture into the nonlinear range and approach global or local collapse as the numerical solution becomes increasingly unstable and thus difficult to reach. There are several commercially available Finite Element Analysis software's such as CSI-SAP2000 and CSI-PERFORM-3D which can be used for the seismic performance evaluation of buildings. Moreover, there is research-based finite element analysis platforms such as OpenSees, RUAUMOKO and the older DRAIN-2D/3D, several of which are now open source.

Research for Earthquake Engineering

Research for earthquake engineering means both field and analytical investigation or experimentation intended for discovery and scientific explanation of earthquake engineering related facts, revision of conventional concepts in the light of new findings, and practical application of the developed theories. The National Science Foundation (NSF) is the main United States government agency that supports fundamental research and education in all fields of earthquake engineering. In particular, it focuses on experimental, analytical, and computational research on design and performance enhancement of structural systems.

The Earthquake Engineering Research Institute (EERI) is a leader in dissemination of earthquake engineering research related information both in the U.S. and globally.

A definitive list of earthquake engineering research related shaking tables around the world may be found in Experimental Facilities for Earthquake Engineering Simulation Worldwide. The most prominent of them is now E-Defence Shake Table in Japan.

Major U.S. Research Programs

The NSF Hazard Mitigation and Structural Engineering program (HMSE) supports research on new technologies for improving the behaviour and response of structural systems subject to earthquake hazards; fundamental research on safety and reliability of constructed systems; innovative developments in analysis and model based simulation of structural behaviour and response including soil-structure interaction; design concepts that improve structure performance and flexibility; and application of new control techniques for structural systems.

NSF also supports the George E. Brown, Jr. Network for Earthquake Engineering Simulation (NEES) that advances knowledge discovery and innovation for earthquakes and tsunami loss reduction of the nation's civil infrastructure, and new experimental simulation techniques and instrumentation. The NEES network features 14 geographically-distributed, shared-use laboratories that support several types of experimental work: geotechnical centrifuge research, shake-table tests, large-scale structural testing, tsunami wave basin experiments, and field site research. Participating universities include: Cornell University; Lehigh University; Oregon State University; Rensselaer Polytechnic Institute; University at Buffalo, SUNY; University of California, Berkeley; University of California, Davis; University of California, Los Angeles; University of California, San Diego; University of California, Santa Barbara; University of Illinois, Urbana-Champaign; University of Minnesota; University of Nevada, Reno; and the University of Texas, Austin.

The equipment sites (labs) and a central data repository are connected to the global earthquake engineering community via the NEEShub website. The NEES website is powered by HUBzero software developed at Purdue University for nanoHUB specifically to help the scientific community share resources and collaborate. The cyberinfrastructure, connected via Internet2, provides interactive simulation tools, a simulation tool development area, a curated central data repository, animated presentations, user support, telepresence, mechanism for uploading and sharing resources and statistics about users, and usage patterns.

This cyberinfrastructure allows researchers to: securely store, organise and share data within a standardised framework in a central location; remotely observe and participate in experiments through the use of synchronized real-time data and video; collaborate with colleagues to facilitate the planning, performance, analysis, and publication of research experiments; and conduct computational and hybrid simulations that may combine the results of multiple distributed experiments and link physical experiments with computer simulations to enable the investigation of overall system performance.

These resources jointly provide the means for collaboration and discovery to improve the seismic design and performance of civil and mechanical infrastructure systems.

Earthquake Simulation

The very first earthquake simulations were performed by statically applying some *horizontal inertia forces* based on scaled peak ground accelerations to a mathematical model of a building. With the further development of computational technologies, static approaches began to give way to dynamic ones. Dynamic experiments on building and non-building structures may be physical, like shake-table testing, or virtual ones. In both cases, to verify a structure's expected seismic performance, some researchers prefer to deal with so called "real time-histories" though the last cannot be "real" for a hypothetical earthquake specified by either a building code or by some particular research requirements. Therefore, there is a strong incentive to engage an earthquake simulation which is the seismic input that possesses only essential features of a real event.

Sometimes, earthquake simulation is understood as a recreation of local effects of a strong earth shaking.

Structure Simulation

Theoretical or experimental evaluation of anticipated seismic performance mostly requires a structure simulation which is based on the concept of structural likeness or similarity. Similarity is some degree of analogy or resemblance between two or more objects. The notion of similarity rests either on exact or approximate repetitions of patterns in the compared items.

In general, a building model is said to have similarity with the real object if the two share *geometric similarity*, *kinematic similarity* and *dynamic similarity*. The most vivid and effective type of similarity is the *kinematic* one. *Kinematic similarity* exists when the paths and velocities of moving particles of a model and its prototype are similar.

The ultimate level of *kinematic similarity* is *kinematic equivalence* when, in the case of earthquake engineering, time-histories of each story lateral displacements of the model and its prototype would be the same.

Seismic Vibration Control

Seismic vibration control is a set of technical means aimed to mitigate seismic impacts in building and non-building structures. All seismic vibration control devices may be classified as *passive, active* or *hybrid* where:

- *passive control devices* have no feedback capability between them, structural elements and the ground;
- *active control devices* incorporate real-time recording instrumentation on the ground integrated with earthquake input processing equipment and actuators within the structure;
- *hybrid control devices* have combined features of active and passive control systems.

When ground seismic waves reach up and start to penetrate a base of a building, their energy flow density, due to reflections, reduces dramatically: usually, up to 90%. However, the remaining portions of the incident waves during a major earthquake still bear a huge devastating potential.

After the seismic waves enter a superstructure, there are a number of ways to control them in order to soothe their damaging effect and improve the building's seismic performance, for instance:

- to dissipate the wave energy inside a superstructure with properly engineered dampers;
- to disperse the wave energy between a wider range of frequencies;
- to absorb the resonant portions of the whole wave frequencies band with the help of so called *mass dampers*.

Figure: *Mausoleum of Cyrus, the oldest base-isolated structure in the world*

Devices of the last kind, abbreviated correspondingly as TMD for the tuned (*passive*), as AMD for the *active*, and as HMD for the *hybrid mass dampers*, have been studied and installed in high-rise buildings, predominantly in Japan, for a quarter of a century.

However, there is quite another approach: partial suppression of the seismic energy flow into the superstructure known as seismic or base isolation.

For this, some pads are inserted into or under all major load-carrying elements in the base of the building which should substantially decouple a superstructure from its substructure resting on a shaking ground.

The first evidence of earthquake protection by using the principle of base isolation was discovered in Pasargadae, a city in ancient Persia, now Iran: it goes back to 6th century BCE. Below, there are some samples of seismic vibration control technologies of today.

Dry-stone Walls Control

People of Inca civilization were masters of the polished *'dry-stone walls'*, called ashlar, where blocks of stone were cut to fit together tightly without any mortar. The Incas were among the best stonemasons the world has ever seen, and many junctions in their masonry were so perfect that even blades of grass could not fit between the stones.

Peru is a highly seismic land, and for centuries the mortar-free construction proved to be apparently more earthquake-resistant than using mortar. The stones of the dry-stone walls built by the Incas could move slightly and resettle without the walls collapsing, a passive structural control technique employing both the principle of energy dissipation and that of suppressing resonant amplifications.

Lead Rubber Bearing

Lead Rubber Bearing or LRB is a type of base isolation employing a heavy damping. It was invented by Bill Robinson, a New Zealander.

Heavy damping mechanism incorporated in vibration control technologies and, particularly, in base isolation devices, is often considered a valuable source of suppressing vibrations thus enhancing a building's seismic performance. However, for the rather pliant systems such as base isolated structures, with a relatively low bearing stiffness but with a high damping, the so-called "damping force" may turn out the main pushing force at a strong earthquake. The video shows a Lead Rubber Bearing being tested at the UCSD Caltrans-SRMD facility. The bearing is made of rubber with a lead core. It was a uniaxial test in which the bearing was also under a full structure load. Many buildings and bridges, both in New Zealand and elsewhere, are protected with lead dampers and lead and rubber bearings. Te Papa Tongarewa, the national museum of New Zealand, and the New Zealand Parliament Buildings have been fitted with the bearings. Both are in Wellington, which sits on an active earthquake fault.

Tuned Mass Damper

Typically, the tuned mass dampers are huge concrete blocks mounted in skyscrapers or other structures and moved in opposition to the resonance frequency oscillations of the structures by means of some sort of spring mechanism. Taipei 101 skyscraper needs to withstand typhoon winds and earthquake tremors common in its area of the Asia-Pacific. For this purpose, a steel pendulum weighing 660 metric tons that serves as a tuned mass damper was designed and installed atop the structure. Suspended from the 92nd to the 88th floor, the pendulums sways to decrease resonant amplifications of lateral displacements in the building caused by earthquakes and strong gusts.

Friction Pendulum Bearing

Friction Pendulum Bearing (FPB) is another name of Friction Pendulum System (FPS). It is based on three pillars:

- articulated friction slider;
- spherical concave sliding surface;
- enclosing cylinder for lateral displacement restraint.

Snapshot with the link to video clip of a shake-table testing of FPB system supporting a rigid building model is presented at the right.

Building Elevation Control

Building elevation control is a valuable source of vibration control of seismic loading. Pyramid-shaped skyscrapers continue to attract attention of architects and engineers because such structures promise a better stability against earthquakes and winds. The elevation configuration can prevent buildings' resonant amplifications because a properly configured building disperses the shear wave energy between a wide range of frequencies.

Earthquake or wind quieting ability of the elevation configuration is provided by a specific pattern of multiple reflections and transmissions of vertically propagating shear waves, which are generated by breakdowns into homogeneity of story layers, and a taper. Any abrupt changes of the propagating waves velocity result in a considerable dispersion of the wave energy between a wide ranges of frequencies thus preventing the resonant displacement amplifications in the building.

A tapered profile of a building is not a compulsory feature of this method of structural control. A similar resonance preventing effect can be also obtained by a proper *tapering* of other characteristics of a building structure, namely, its mass and stiffness. As a result, the building elevation configuration techniques permit an architectural design that may be both attractive and functional.

Simple Roller Bearing

Simple roller bearing is a base isolation device which is intended for protection of various building and non-building structures against potentially damaging lateral impacts of strong earthquakes. This metallic bearing support may be adapted, with certain precautions, as a seismic isolator to skyscrapers and buildings on soft ground. Recently, it has been employed under the name of *Metallic Roller Bearing* for a housing complex (17 stories) in Tokyo, Japan.

Springs-with-damper Base Isolator

Springs-with-damper base isolator installed under a three-story town-house, Santa Monica, California is shown on the photo taken prior to the 1994 Northridge earthquake exposure. It is a base isolation device conceptually similar to *Lead Rubber Bearing.*

One of two three-story town-houses like this, which was well instrumented for recording of both vertical and horizontal accelerations on its floors and the ground, has survived a severe shaking during the Northridge earthquake and left valuable recorded information for further study.

Hysteretic Damper

Hysteretic damper is intended to provide better and more reliable seismic performance than that of a conventional structure at the expense of the seismic input energy dissipation. There are four major groups of hysteretic dampers used for the purpose, namely:

- Fluid viscous dampers (FVDs)
- Metallic yielding dampers (MYDs)
- Viscoelastic dampers (VEDs)
- Friction dampers (FDs).

Each group of dampers has specific characteristics, advantages and disadvantages for structural applications.

Seismic Design

Seismic design is based on authorised engineering procedures, principles and criteria meant to design or retrofit structures subject to earthquake exposure. Those criteria are only consistent with the contemporary state of the knowledge about earthquake engineering structures. Therefore, a building design which exactly follows seismic code regulations does not guarantee safety against collapse or serious damage. The price of poor seismic design may be enormous. Nevertheless, seismic design has always been a trial and error process whether it was based on physical laws or on empirical knowledge of the structural performance of different shapes and materials.

To practice seismic design, seismic analysis or seismic evaluation of new and existing civil engineering projects, an engineer should, normally, pass examination on *Seismic Principles* which, in the State of California, include:

- Seismic Data and Seismic Design Criteria
- Seismic Characteristics of Engineered Systems
- Seismic Forces
- Seismic Analysis Procedures
- Seismic Detailing and Construction Quality Control.

To build up complex structural systems, seismic design largely uses the same relatively small number of basic structural elements (to say nothing of vibration control devices) as any non-seismic design project.

Normally, according to building codes, structures are designed to "withstand" the largest earthquake of a certain probability that is likely to occur at their location. This means the loss of life should be minimised by preventing collapse of the buildings.

Seismic design is carried out by understanding the possible failure modes of a structure and providing the structure with appropriate strength, stiffness, ductility, and configuration to ensure those modes cannot occur.

Seismic Design Requirements

Seismic design requirements depend on the type of the structure, locality of the project and its authorities which stipulate applicable seismic design codes and criteria. For instance, California Department of Transportation's requirements called *The Seismic Design Criteria* (SDC) and aimed at the design of new bridges in California incorporate an innovative seismic performance based approach.

The most significant feature in the SDC design philosophy is a shift from a *force-based assessment* of seismic demand to a *displacement-based assessment* of demand and capacity. Thus, the newly adopted displacement approach is based on comparing the *elastic displacement* demand to the *inelastic displacement* capacity of the primary structural components while ensuring a minimum level of inelastic capacity at all potential plastic hinge locations.

In addition to the designed structure itself, seismic design requirements may include a *ground stabilisation* underneath the structure: sometimes, heavily shaken ground breaks up which leads to collapse of the structure sitting upon it. The following topics should be of primary concerns: liquefaction; dynamic lateral earth pressures on retaining walls; seismic slope stability; earthquake-induced settlement.

Nuclear facilities should not jeopardise their safety in case of earthquakes or other hostile external events. Therefore, their seismic design is based on criteria far more stringent than those applying to non-nuclear facilities. The Fukushima I nuclear accidents and damage to other nuclear facilities that followed the 2011 Tôhoku earthquake and tsunami have, however, drawn attention to ongoing concerns over Japanese nuclear seismic design standards and caused other many governments to re-evaluate their nuclear programs. Doubt has also been expressed over the seismic evaluation and design of certain other plants, including the Fessenheim Nuclear Power Plant in France.

Failure Modes

Failure mode is the manner by which an earthquake induced failure is observed. It, generally, describes the way the failure occurs. Though costly and time consuming, learning from each real earthquake failure remains a routine recipe for advancement in *seismic design* methods. Below, some typical modes of earthquake-generated failures are presented. The lack of reinforcement coupled with poor mortar and inadequate roof-to-wall ties can result in substantial damage to a unreinforced masonry building. Severely cracked or leaning walls are some of the most common earthquake damage. Also hazardous is the damage that may occur between the walls and roof or floor diaphragms. Separation between the framing and the walls can jeopardize the vertical support of roof and floor systems.

Soft story effect. Absence of adequate shear walls on the ground level caused damage to this structure. A close examination of the image reveals that the rough board siding, once covered by a brick veneer, has been completely dismantled from the studwall. Only the rigidity of the floor above combined with the support on the two hidden sides by continuous walls, not penetrated with large doors as on the street sides, is preventing full collapse of the structure.

Soil liquefaction. In the cases where the soil consists of loose granular deposited materials with the tendency to develop excessive hydrostatic pore water pressure of sufficient magnitude and compact, liquefaction of those loose saturated deposits may result in non-uniform settlements and tilting of structures. This caused major damage to thousands of buildings in Niigata, Japan during the 1964 earthquake.

Landslide rock fall. A landslide is a geological phenomenon which includes a wide range of ground movement, including rock falls. Typically, the action of gravity is the primary driving force for a landslide to occur though in this case there was another contributing factor which affected the original slope stability: the landslide required an *earthquake trigger* before being released.

Pounding against adjacent building. This is a photograph of the collapsed five-story tower, St. Joseph's Seminary, Los Altos, California which resulted in one fatality. During Loma Prieta earthquake, the tower pounded against the independently vibrating adjacent building behind. A possibility of pounding depends on both buildings' lateral displacements which should be accurately estimated and accounted for.

At Northridge earthquake, the Kaiser Permanente concrete frame office building had joints completely shattered, revealing inadequate confinement steel, which resulted in the second story collapse. In the transverse direction, composite end shear walls, consisting of two wythes of brick and a layer of shotcrete that carried the lateral load, peeled apart because of inadequate through-ties and failed.

7-story reinforced concrete buildings on steep slope collapse due to the following :

- Improper construction site on a foothill.
- Poor detailing of the reinforcement (lack of concrete confinement in the columns and at the beam-column joints, inadequate splice length).
- Seismically weak soft story at the first floor.
- Long cantilevers with heavy dead load.

Sliding off foundations effect of a relatively rigid residential building structure during 1987 Whittier Narrows earthquake. The magnitude 5.9 earthquake pounded the Garvey West Apartment building in Monterey Park, California and shifted its superstructure about 10 inches to the east on its foundation. If a superstructure is not mounted on a base isolation system, its shifting on the basement should be prevented.

Reinforced concrete column burst at Northridge earthquake due to insufficient shear reinforcement mode which allows main reinforcement to buckleoutwards. The deck unseated at the hinge and failed in shear. As a result, the La Cienega-Venice underpass section of the 10 Freeway collapsed. Loma Prieta earthquake: side view of reinforced concrete support-columns failure which trigged the upper deck collapse onto the lower deck of the two-level Cypress viaduct of Interstate Highway 880, Oakland, CA.

Retaining wall failure at Loma Prieta earthquake in Santa Cruz Mountains area: prominent northwest-trending extensional cracks up to 12 cm (4.7 in) wide in the concrete spillway to Austrian Dam, the north abutment. Ground shaking triggered soil liquefaction in a subsurface layer of sand, producing differential lateral and vertical movement in an overlying carapace of unliquified sand and silt. This mode of ground failure, termed lateral spreading, is a principal cause of liquefaction-related earthquake damage.

Severely damaged building of Agriculture Development Bank of China after 2008 Sichuan earthquake: most of the beams and pier columns are sheared. Large diagonal cracks in masonry and veneer are due to in-plane loads while abrupt settlement of the right end of the building should be attributed to a landfill which may be hazardous even without any earthquake. Twofold tsunami impact: sea waves hydraulic pressure and inundation. Thus, 2004 Indian Ocean earthquake of December 26, 2004, with the epicentre off the west coast of Sumatra, Indonesia, triggered a series of devastating tsunamis, killing more than 225,000 people in eleven countries by inundating surrounding coastal communities with huge waves up to 30 metres (100 feet) high.

Earthquake-Resistant Construction

Earthquake construction means implementation of seismic design to enable building and non-building structures to live through the anticipated earthquake exposure up to the expectations and in compliance with the applicable building codes.

Design and construction are intimately related. To achieve a good workmanship, detailing of the members and their connections should be, possibly, simple. As any construction in general, earthquake construction is a process that consists of the building, retrofitting or assembling of infrastructure given the construction materials available.

The destabilising action of an earthquake on constructions may be *direct* (seismic motion of the ground) or *indirect* (earthquake-induced landslides, soil liquefaction and waves of tsunami).

A structure might have all the appearances of stability, yet offer nothing but danger when an earthquake occurs. The crucial fact is that, for safety, earthquake-resistant construction techniques are as important as quality control and using correct materials. *Earthquake contractor* should be registered in the state of the project location, bonded and insured.

To minimise possible losses, construction process should be organised with keeping in mind that earthquake may strike any time prior to the end of construction.

Each construction project requires a qualified team of professionals who understand the basic features of seismic performance of different structures as well as construction management.

Adobe Structures

Around thirty percent of the world's population lives or works in earth-made construction. Adobe type of mud bricks is one of the oldest and most widely used building materials. The use of adobe is very common in some of the world's most hazard-prone regions, traditionally across Latin America, Africa, Indian subcontinent and other parts of Asia, Middle East and Southern Europe. Adobe buildings are considered very vulnerable at strong quakes. However, multiple ways of seismic strengthening of new and existing adobe buildings are available, Key factors for the improved seismic performance of adobe construction are:

- Quality of construction.
- Compact, box-type layout.
- Seismic reinforcement.

Limestone and Sandstone Structures

Figure: *Base-isolated City and County Building, Salt Lake City, Utah*

Limestone is very common in architecture, especially in North America and Europe. Many landmarks across the world, including the pyramids in Egypt, are made of limestone. Many medieval churches and castles in Europe are made of limestone and sandstone masonry. They are the long-lasting materials but their rather heavy weight is not beneficial for adequate seismic performance.

Application of modern technology to seismic retrofitting can enhance the survivability of unreinforced masonry structures. As an example, from 1973 to 1989, the Salt Lake City and County Building in Utah was exhaustively renovated and repaired with an emphasis on preserving historical accuracy in appearance. This was done in concert with a seismic upgrade that placed the weak sandstone structure on base isolation foundation to better protect it from earthquake damage.

Timber Frame Structures

Timber framing dates back thousands of years, and has been used in many parts of the world during various periods such as ancient Japan, Europe and medieval England in localities where timber was in good supply and building stone and the skills to work it were not.

Figure: *Half-timbered museum buildings,Denmark, date from 1560*

The use of timber framing in buildings provides their complete skeletal framing which offers some structural benefits as the timber frame, if properly engineered, lends itself to better *seismic survivability*.

Light-frame Structures

Light-frame structures usually gain seismic resistance from rigid plywood shear walls and wood structural panel diaphragms. Special provisions for seismic load-resisting systems for all engineered wood structures requires consideration of diaphragm ratios, horizontal and vertical diaphragm shears, and connector/fastener values. In addition, collectors, or drag struts, to distribute shear along a diaphragm length are required.

Reinforced Masonry Structures

A construction system where steel reinforcement is embedded in the mortar joints of masonry or placed in holes and after filled with concrete or grout is called reinforced masonry.

The devastating 1933 Long Beach earthquake revealed that masonry construction should be improved immediately. Then, the California State Code made the reinforced masonry mandatory.

There are various practices and techniques to achieve reinforced masonry. The most common type is the reinforced hollow unit masonry. The effectiveness of both vertical and horizontal reinforcement strongly depends on the type and quality of the masonry, i.e. masonry units and mortar.

To achieve a ductile behaviour of masonry, it is necessary that the shear strength of the wall is greater than the tensile strength of reinforcement to ensure a kind of bending failure.

Reinforced Concrete Structures

Reinforced concrete is concrete in which steel reinforcement bars (rebars) or fibres have been incorporated to strengthen a material that would otherwise bebrittle. It can be used to produce beams, columns, floors or bridges.

Prestressed concrete is a kind of reinforced concrete used for overcoming concrete's natural weakness in tension. It can be applied to beams, floors or bridges with a longer span than is practical with ordinary reinforced concrete. Prestressing tendons (generally of high tensile steel cable or rods) are used to provide a clamping load which produces a compressive stress that offsets the tensile stress that the concrete compression member would, otherwise, experience due to a bending load.

To prevent catastrophic collapse in response earth shaking (in the interest of life safety), a traditional reinforced concrete frame should have ductile joints. Depending upon the methods used and the imposed seismic forces, such buildings may be immediately usable, require extensive repair, or may have to be demolished.

Prestressed Structures

Prestressed structure is the one whose overall integrity, stability and security depend, primarily, on a *prestressing*. *Prestressing* means the intentional creation of permanent stresses in a structure for the purpose of improving its performance under various service conditions.

There are the following basic types of prestressing:

- Pre-compression (mostly, with the own weight of a structure)
- Pretensioning with high-strength embedded tendons
- Post-tensioning with high-strength bonded or unbonded tendons.

Today, the concept of prestressed structure is widely engaged in design of buildings, underground structures, TV towers, power stations, floating storage and offshore facilities, nuclear reactor vessels, and numerous kinds of bridge systems.

A beneficial idea of *prestressing* was, apparently, familiar to the ancient Rome architects; look, e.g., at the tall attic wall of Colosseum working as a stabilising device for the wall piers beneath.

Steel Structures

Figure: *Collapsed section of the San Francisco - Oakland Bay Bridge in response to Loma Prieta earthquake*

Steel structures are considered mostly earthquake resistant but their resistance should never be taken for granted. A great number of welded steel moment frame buildings, which looked earthquake-proof, surprisingly experienced brittle behaviour and were hazardously damaged in the 1994 Northridge earthquake. After that, the Federal Emergency Management Agency (FEMA) initiated development of repair techniques and new design approaches to minimise damage to steel moment frame buildings in future earthquakes.

For structural steel seismic design based on Load and Resistance Factor Design (LRFD) approach, it is very important to assess ability of a structure to develop and maintain its bearing resistance in the inelastic range. A measure of this ability is ductility, which may be observed in a *material itself*, in a structural *element*, or to a *whole structure*.

As a consequence of Northridge earthquake experience, all pre-qualified connection details and design methods contained in the building codes of that time have been rescinded. The new provisions stipulated that new designs be substantiated by testing or by use of test-verified calculations.

Prediction of Earthquake Losses

Earthquake loss estimation is usually defined as a *Damage Ratio* (DR) which is a ratio of the earthquake damage repair cost to the total value of a building. *Probable Maximum Loss* (PML) is a common term used for earthquake loss estimation, but it lacks a precise definition. In 1999, ASTM E2026 'Standard Guide for the Estimation of Building Damageability in Earthquakes' was produced in order to standardise the nomenclature for seismic loss estimation, as well as establish guidelines as to the review process and qualifications of the reviewer.

Earthquake loss estimations are also referred to as *Seismic Risk Assessments*. The risk assessment process generally involves determining the probability of various ground motions coupled with the vulnerability or damage of the building under those ground motions. The results are defined as a percent of building replacement value.

Probabilistic Risk Assessment

Probabilistic risk assessment (PRA) is a systematic and comprehensive methodology to evaluate risks associated with a complex engineered technological entity (such as an airliner or a nuclear power plant).

Risk in a PRA is defined as a feasible detrimental outcome of an activity or action. In a PRA, risk is characterised by two quantities:

1. the magnitude (severity) of the possible adverse consequence(s), and
2. the likelihood (probability) of occurrence of each consequence.

Consequences are expressed numerically (e.g., the number of people potentially hurt or killed) and their likelihoods of occurrence are expressed as probabilities or frequencies (i.e., the number of occurrences or the probability of occurrence per unit time). The total risk is the expected loss: the sum of the products of the consequences multiplied by their probabilities.

The spectrum of risks across classes of events are also of concern, and are usually controlled in licensing processes – it would be of concern if rare but high consequence events were found to dominate the overall risk, particularly as these risk assessment is very sensitive to assumptions (how rare is a high consequence event?).

Probabilistic Risk Assessment usually answers three basic questions:

1. What can go wrong with the studied technological entity, or what are the initiators or initiating events (undesirable starting events) that lead to adverse consequence(s)?
2. What and how severe are the potential detriments, or the adverse consequences that the technological entity may be eventually subjected to as a result of the occurrence of the initiator?
3. How likely to occur are these undesirable consequences, or what are their probabilities or frequencies?

Two common methods of answering this last question are Event Tree Analysis and Fault Tree Analysis.

In addition to the above methods, PRA studies require special but often very important analysis tools like human reliability analysis (HRA) and common-cause-failure analysis (CCF). HRA deals with methods for modelling human error while CCF deals with methods for evaluating the effect of inter-system and intra-system dependencies which tend to cause simultaneous failures and thus significant increases in overall risk.

In 2007 France was criticised for failing to use a PRA approach to evaluate the seismic risks of French nuclear power plants.

Criticism

Theoretically, the probabilistic risk assessment method suffers from several problems:

These risk assessments do a poor job of modelling human actions and their impact on known, let alone unknown, failure modes. Also, as a 1978 Risk Assessment Review Group Report to the NRC pointed out, it is "conceptually impossible to be complete in a mathematical sense in the construction of event-trees and fault-trees ... This inherent limitation means that any calculation using this methodology is always subject to revision and to doubt as to its completeness."

In the case of many accidents, probabilistic risk assessment models do not account for unexpected failure modes:

At Japan's Kashiwazaki Kariwa reactors, for example, after the 2007 Chuetsu earthquake some radioactive materials escaped into the sea when ground subsidence pulled underground electric cables downward and created an opening in the reactor's basement wall. As a Tokyo Electric Power Company official remarked then, "It was beyond our imagination that a space could be made in the hole on the outer wall for the electric cables."

When it comes to future safety, nuclear designers and operators often assume that they know what is likely to happen, which is what allows them to assert that they have planned for all possible contingencies. Yet there is one weakness of the probabilistic risk assessment method that has been emphatically demonstrated with the Fukushima I nuclear accidents — the difficulty of modelling common-cause or common-mode failures:

From most reports it seems clear that a single event, the tsunami, resulted in a number of failures that set the stage for the accidents. These failures included the loss of offsite electrical power to the reactor complex, the loss of oil tanks and replacement fuel for diesel generators, the flooding of the electrical switchyard, and perhaps damage to the inlets that brought in cooling water from the ocean. As a result, even though there were multiple ways of removing heat from the core, all of them failed.

Soil Structure Interaction

Most of the civil engineering structures involve some type of structural element with direct contact with ground. When the external forces, such as earthquakes, act on these systems, neither the structural displacements nor the ground displacements, are independent of each other. The process in which the response of the soil influences the motion of the structure and the motion of the structure influences the response of the soil is termed as soil-structure interaction (SSI).

Conventional structural design methods neglect the SSI effects. Neglecting SSI is reasonable for light structures in relatively stiff soil such as low rise buildings and simple rigid retaining walls. The effect of SSI, however, becomes prominent for heavy structures resting on relatively soft soils for example nuclear power plants, high-rise buildings and elevated-highways on soft soil.

Damage sustained in recent earthquakes, such as the 1995 Kobe Earthquake, have also highlighted that the seismic behaviour of a structure is highly influenced not only by the response of the superstructure, but also by the response of the foundation and the ground as well . Hence, the modern seismic design codes, such as Standard Specifications for Concrete Structures: Seismic Performance Verification JSCE 2005 stipulate that the response analysis should be conducted by taking into consideration a whole structural system including superstructure, foundation and ground.

Effect of Soil Structure Interaction on Structural Response

It has conventionally been considered that soil-structure interaction has beneficial effect on the seismic response of a structure. Many design codes have suggested that the effect of SSI can reasonably be neglected for the seismic analysis of structures . This myth about SSI apparently stems from the false perception that SSI reduces the overall seismic response of a structure, and hence, leads to improved safety margins. Most of the design codes use oversimplified design spectrums, which attain constant acceleration up to a certain period, and thereafter decreases monotonically with period. Considering soil-structure interaction makes a structure more flexible and thus, increasing the natural period of the structure compared to the corresponding rigidly supported structure. Moreover, considering the SSI effect increases the effective damping ratio of the system. The smooth idealisation of design spectrum suggests smaller seismic response with the increased natural periods and effective damping ratio due to SSI. With this assumption, it was traditionally been considered that SSI can conveniently be neglected for conservative design. In addition, neglecting SSI tremendously reduces the complication in the analysis of the structures which has tempted designers to neglect the effect of SSI in the analysis. This conservative simplification is valid for certain class of structures and soil conditions, such as light structures in relatively stiff soil. Unfortunately, the assumption does not always hold true. In fact, the SSI can have a detrimental effect on the structural response, and neglecting SSI in the analysis may lead to unsafe design for both the superstructure and the foundation .

Detrimental Effects of SSI

Using rigorous numerical analyses, Mylonakis and Gazetas have shown that increase in natural period of structure due to SSI is not always beneficial as suggested by the simplified design spectrums. Soft soil sediments can significantly elongate the period of seismic waves and the increase in natural period of structure may lead to the resonance with the long period ground vibration. Additionally, the study showed that ductility demand can significantly increase with the increase in the natural period of the structure due to SSI effect. The permanent deformation and failure of soil may further aggravate the seismic response of the structure.

When a structure is subjected to an earthquake excitation, it interacts with the foundation and the soil, and thus changes the motion of the ground. Soil-structure interaction broadly can be divided into two phenomena: a) kinematic interaction and b) inertial interaction. Earthquake ground motion causes soil displacement known as free-field motion. However, the foundation embedded into the soil will not follow the free field motion. This inability of the foundation to match the free field motion causes the kinematic interaction. On the other hand, the mass of the super-structure transmits the inertial force to the soil causing further deformation in the soil, which is termed as inertial interaction. At low level of ground shaking, kinematic effect is more dominant causing the lengthening of period and increase in radiation damping. However, with the onset of stronger shaking, near-field soil modulus degradation and soil-pile gapping limit radiation damping, and inertial interaction becomes predominant causing excessive displacements and bending strains concentrated near the ground surface resulting in pile damage near the ground level .

Observations from recent earthquakes have shown that the response of the foundation and soil can greatly influence the overall structural response. There are several cases of severe damages in structures due to SSI in the past earthquakes. Yashinsky cites damage in number of pile-supported bridge structures due to SSI effect in Loma Prieta Earthquake in San Francisco in 1989. Extensive numerical analysis carried out by Mylonakis and Gazetas have attributed SSI as one of the reasons behind the dramatic collapse of Hanshin Expressway in 1995 Kobe Earthquake.

Spectral Acceleration

Spectral acceleration (SA) is a unit measured in *g* (the acceleration due to Earth's gravity, equivalent to g-force) that describes the maximum acceleration in an earthquakeon an object – specifically a damped, harmonic oscillator moving in one physical dimension. This can be measured at (or specified for) different oscillation frequencies and with different degrees of damping, although 5% damping is commonly applied. The SA at different frequencies may be plotted to form a response spectrum. Spectral acceleration, with a value related to the natural frequency of vibration of the building, is used in earthquake engineering and gives a closer approximation to the motion of a building or other structure in an earthquake than the peak ground acceleration value, although there is normally a correlation between SA and PGA.

Base Isolation

Base isolation, also known as seismic base isolation or base isolation system, is one of the most popular means of protecting a structure against earthquake forces. It is a collection of structural elements which should substantially decouple a superstructure from its substructure resting on a shaking ground thus protecting a building or non-building structure's integrity. Base isolation is the most powerful tool of earthquake engineering pertaining to the passive structural vibration control technologies. It is meant to enable a building or non-building structure to survive a potentially devastating seismic impact through a proper initial design or subsequent modifications. In some cases, application of base isolation can raise both a structure's seismic performance and its seismic sustainability considerably. Contrary to popular belief base isolation does not make a building earthquake proof.

Base isolation system consists of *isolation units* with or without *isolation components*, where:

1. *Isolation units* are the basic elements of *base isolation system* which are intended to provide the mentioned decoupling effect to a building or non-building structure.
2. *Isolation components* are the connections between *isolation units* and their parts having no decoupling effect of their own.

By their response to an earthquake impact, all *isolation units* may be divided into two basic categories: *shear units* and *sliding units*. The first evidence of architects using the principle of base isolation for earthquake protection was discovered in Pasargadae , a city in ancient Persia, now Iran: it goes back to 6th century BC. It works by having a wide and deep stone and mortar foundation, smoothed at the top, upon which a second foundation is built of wide, smoothed stones which are linked together, forming a plate that slides back and forth over the lower foundation in case of an earthquake leaving the structure intact.

This technology can be used both for new structural design and seismic retrofit. In process of seismic retrofit, some of the most prominent U.S. monuments, e.g. Pasadena City Hall, San Francisco City Hall, Salt Lake City and County Building or LA City Hall were mounted on *Base Isolation Systems*. It required creating rigidity diaphragms and moats around the buildings, as well as making provisions against overturning and P-Delta Effect.

2

Geotechnical-engineering Specialities

Although geotechnical engineering is applied for a variety of purposes, it is essential to foundation design. As such, geotechnical engineering is applicable to every existing or new structure on the planet; every building and every highway, bridge, tunnel, harbor, airport, water line, reservoir, or other public work. Commonly, the geotechnical-engineering service comprises a study of subsurface conditions using various sampling, in-situ testing, and/or other site-characterisation techniques.

The instrument of professional service in those cases typically is a report through which geotechnical engineers relate the information they have been retained to provide, typically: their findings; their opinions about subsurface materials and conditions; their judgment about how the subsurface materials and conditions assumed to exist probably will behave when subjected to loads or used as building material; and their preliminary recommendations for materials usage or appropriate foundation systems, the latter based on their knowledge of a structure's size, shape, weight, etc., and the subsurface/structure interactions likely to occur. Civil engineers, structural engineers, and architects, feasibly among other members of the project team, apply the geotechnical findings and preliminary recommendations to take the structure's design forward. They realise these preliminary recommendations are subject to change, however, because – as a matter of practical necessity related to the observational method inherent to geotechnical engineering – geotechnical engineers base their recommendations on the composition of samples taken from a tiny portion of a site whose actual subsurface conditions are unknowable before excavation, because they are hidden by earth and/or rock and/or water. For this reason, as a key component of a complete geotechnical engineering service, geotechnical engineers employ construction-materials engineering and testing (CoMET) to observe subsurface materials as they are exposed through excavation. To help achieve economies on their clients' behalf, geotechnical engineers assign their field representatives – specially educated and trained paraprofessionals– to observe the excavated materials and the excavations themselves in light of conditions the geotechnical engineers opined to exist. When differences are discovered, the geotechnical engineers evaluate the new findings and, when necessary, modify their design and construction recommendations. Because such changes could require other members of the design and construction team to modify their designs, specifications, and proposed methods, many owners have their geotechnical engineers serve as active members of the project team from project inception to conclusion, working with others to help ensure appropriate application of geotechnical information and judgments.

In other cases, geotechnical engineering goes beyond a study and construction recommendations to include design of soil and rock structures. The most common of these are the pavements that make up our streets and highways, airport runways, and bridge and tunnel decks, among other paved improvements. Geotechnical engineers design the pavements in terms of the subgrade, subbase, and base layers of materials to be used, and the thickness and composition of each. Geotechnical engineers also design the earth-retention walls associated with structures such as levees, earthen dams, reservoirs, and landfills. In other cases, the design is applied to contain earth, via structures such as excavation-support systems and retaining walls. Sometimes referred to as geostructural engineering or geostructural design, these services are also intrinsic to hydraulic engineering, hydrogeologic engineering, coastal engineering, geologic engineering and water-resources engineering. Geotechnical-engineering design is also applied for structures such as tunnels, bridges, dams, and other structures beneath, on, or connected to the surface of the earth. Geotechnical engineering, like geology, engineering geology, and geologic engineering, also involves the specialities of rock mechanics and soil mechanics, and often requires knowledge of geotextiles and geosynthetics, as well as an array of instrumentation and monitoring equipment, to help ensure specified conditions are achieved and maintained.

Earthquake engineering and landslide detection, remediation, and prevention are geoprofessional services associated with specialised types of geotechnical engineering, as is forensic geotechnical engineering, a geoprofessional service applied to determine why a certain applicable type of event – usually a failure of some sort – occurred. (Virtually all geoprofessional services can be performed for forensic purposes, commonly as litigation-support/expert witness services.)

Railway-systems engineering is another type of specialised geotechnical engineering, as are the design of piers and bulkheads, drydocks, on-shore and off-shore wind-turbine systems, and systems that stabilise oil platforms and other marine structures to the sea floor. Geotechnical engineers have long been involved in sustainability initiatives, including (among many others) the use of excavated materials; the safe application of contaminated subsurface materials; the recycling of asphalt, concrete, and building rubble and debris; and the design of permeable pavements.

All civil-engineering specialities and projects – roads and highways, bridges, rail systems, ports and other waterfront structures, airport terminals, etc. – require the involvement of geotechnical engineers and engineering, meaning that many civil-engineering pursuits are geoprofessional pursuits to a greater or lesser degree. However, geotechnical engineering has for centuries also been associated with military engineering; sappers (in general) and miners (whose tunnelling design services (known as landmining and undermining) were used in military-siege operations).

Engineering Geology and other Geology Specialities

Engineering geologist. (a) Elements of the engineering geologist speciality. The practice of engineering geology involves the interpretation, evaluation, analysis, and application of geological information and data to civil works. Geotechnical soil and rock units are designated, characterised, and classified, using standard engineering soil and rock classification systems. Relationships are interpreted between landform development, current and past geologic processes, ground and surface water, and the strength characteristics of soil and rock. Processes evaluated include both surficial processes (for example, slope, fluvial, and coastal processes), and deep-seated processes (for example, volcanic activity and seismicity). Geotechnical zones or domains are designated based on soil and rock strength characteristics, common landforms, related geologic processes, or other pertinent factors. Proposed developmental modifications are evaluated and, where appropriate, analysed to predict potential or likely changes in types and rates of surficial geologic processes. Proposed modifications may include such things as vegetation removal, using various types of earth materials in construction, applying loads to shallow or deep foundations, constructing cut or fill slopes and other grading, and modifying ground and surface water flow. The effects of surficial and deep-seated geologic processes are evaluated and analysed to predict their potential effect on public health, public safety, land use, or proposed development. (b) Typical engineering geologic applications and types of projects. Engineering geology is applied during all project phases, from conception through planning, design, construction, maintenance, and, in some cases, reclamation and closure.

Planning-level engineering geologic work is commonly conducted in response to forest practice regulations, critical areas ordinances, and the State Environmental Policy Act. Typical planning-level engineering geologic applications include timber harvest planning, proposed location of residential and commercial developments and other buildings and facilities, and alternative route selection for roads, rail lines, trails, and utilities. Site-specific engineering geologic applications include cuts, fills, and tunnels for roads, trails, railroads, and utility lines; foundations for bridges and other drainage structures, retaining walls and shoring, dams, buildings, water towers, slope, channel and shoreline stabilisation facilities, fish ladders and hatcheries, ski lifts and other structures; landings for logging and other work platforms; airport landing strips; rock bolt systems; blasting; and other major earthwork projects such as for aggregate sources and landfills.

While engineering geology is applicable principally to planning, design and construction activities, other specialities of geology are applied in a variety of geoprofessional speciality fields, such as mining geology, petroleum geology, and environmental geology. Note, however, that, while mining geology and mining engineering both are geoprofessional pursuits, they are nonetheless distinctly different from one another.

Geological Engineering

Geological engineering is a hybrid discipline that comprises elements of civil engineering, mining/tunnelling engineering, and engineering geology. Accredited graduate and/or undergraduate geological-engineering programs are offered by about a dozen U.S. schools, including the Colourado School of Mines, the University of

Arizona, and the Missouri University of Science and Technology (formerly University of Missouri-Rolla). Geological engineers often become licensed as both engineers and geologists.

Geophysics

Geophysics is a geoprofessional speciality applied in particular to earthquake engineering and vibration analysis. Vibration analysis is a key element of construction when the project involves techniques such as pile-driving. It also is essential to the effective design, construction, and operation of subterranean and overhead urban mass-transit systems.

Environmental-science and Environmental-engineering Specialities

Environmental science and environmental engineering are the geoprofessions commonly associated with the identification, remediation, and prevention of environmental contamination. These services range from phase-one and phase-two environmental site-assessments – research designed to assess the likelihood that a property is contaminated and subsurface exploration conducted to identify the nature and extent of contamination, respectively – up through the design of processes and systems to remediate contaminated sites for the protection of human health and the environment.

Environmental geology is one of the principal geoprofessions engaged in assessing and remediating contaminated sites. Environmental geologists help identify the subsurface stratigraphy in which contaminants are located and through which they migrate. Environmental chemistry is the geoprofession that encompasses the study of chemical compounds in the soil.

These compounds are categorised as pollutants or contaminants when introduced into the environment by human factors (e.g., waste, mining processes, radioactive release) and are not of natural origin. Environmental chemistry assesses interactions or these compounds with soil, rock, and water to determine their fate and transport, the techniques to measure the levels of contaminants in the environment, and technologies to destroy or reduce the toxicity of contaminants in wastes or compounds that have been released to the environment. Environmental engineering is often applied to assess contaminated sites, but more often is used in the design of systems to remediate contaminated soil and groundwater. Hydrogeology is the geoprofession involved when environmental studies involve subsurface water. Hydrogeology applications range from securing safe, plentiful underground drinking-water sources to identifying the nature of groundwater contamination in order to facilitate remediation.

Environmental toxicology is a geoprofession when used to identify the source, fate, transformation, effects, and risks of pollutants on the environment, including soil, water, and air. Wetlands science is a geoprofessional pursuit that incorporates several scientific disciplines, such as botany, biology, and limnology. It involves, among other activities, the delineation, conservation, restoration, and preservation of wetlands. These services are sometimes conducted by geoprofessional specialists called wetlands scientists. Ecology is a closely related environmental geoprofession involving studies into the distribution of organisms and biodiversity within an environmental context.

Numerous geoprofessional disciplines contribute to the redevelopment of brownfields, sites (typically urban) that are underused or abandoned because they are or are assumed to be contaminated by hazardous materials. Geoprofessionals are engaged to evaluate the degree to which such sites are contaminated and the steps that can be taken to achieve the sites' safe reuse. Environmental engineers and scientists work with developers to identify and design remediation strategies and exposure-barrier designs that protect future site users from unacceptable exposure to environmental contamination resulting from previous uses of the site. Because these previous uses often resulted in degraded soil conditions and the presence of abandoned, underground structures, geotechnical engineers often are needed to design special foundations for the new structures.

Construction-materials Engineering and Testing (CoMET)

Construction-materials engineering and testing (CoMET) comprises an array of licensed-engineer-directed professional services applied principally for purposes of construction quality assurance and quality control. CoMET services commonly are provided as a separate discipline by firms that also practice geotechnical engineering, possibly among other geoprofessional disciplines. The geoprofessional-service industry has evolved in this manner because geotechnical engineering employs the observational method. Karl von Terzaghi and Ralph B. Peck – the creators of modern geotechnical engineering – used the observational method and multiple working hypotheses to expedite and economize the subsurface-exploration process, by using sampling and testing to form a judgment about subsurface conditions, and then observing excavated conditions and materials to confirm or modify those judgments and related recommendations, and then finalise them. To economize still further, geoprofessionals educated and trained paraprofessionals to represent them on site (hence the term "field representative"), especially to apply their judgment (much as a geotechnical engineer would) in comparing observed conditions with those the geotechnical engineer believed would exist. Over time, geotechnical engineers expanded their CoMET services by providing the additional education and training their field representatives needed to evaluate constructors' attainment of conditions commonly specified by geoprofessionals; e.g., subsurface preparation for foundations of buildings, roadways, and other structures; materials used for subgrade, subbase, and base purposes; site grading; construction of earthen structures (earth dams, levees, reservoirs, landfills, et al.) and earth-retaining structures (e.g., retaining walls); and so on. Because many of the materials involved, such as concrete, are used in other elements of construction projects and structures, geoprofessional firms expanded their field representatives' skill sets still more, to encompass observation and testing of numerous additional materials (e.g., reinforced concrete, structural steel, masonry, wood, and fireproofing), processes (e.g., cutting and filling and rebar placement), and outcomes (e.g., the effectiveness of welds). Laboratory services are a common element of many CoMET operations. Also operating under the direction of a licensed engineer, they are applied in geotechnical engineering to evaluate subsurface-material samples. In overall CoMET operations, laboratories operate with the equipment and personnel required to evaluate a variety of construction materials.

CoMET services applied to evaluate the actual composition of a site's subsurface are part of a complete geotechnical engineering service. For purposes of short-term economy, however, some owners select a firm not associated with the geotechnical engineer of record to provide these and all other CoMET services. This approach precludes the geotechnical engineer of record from providing a complete service. It also aggravates risk, because the individuals engaged to evaluate actual subsurface conditions are not "briefed" by the geotechnical engineer of record before they go to the project site and seldom communicate with the geotechnical engineer of record when they discern differences, in large part because the firm associated with the geotechnical engineer of record is regarded as a competitor of the firm employing the field representatives. In some cases, the field representatives in question lack the specific project background information and/or the education and training required to discern those differences.

CoMET services applied to evaluate constructor's attainment of specified conditions take the form of quality-assurance (QA) or quality-control (QC) services. QA services are performed directly or indirectly for the owner. The owner specifies the nature and extent of QA services that the owner believes is appropriate. Some owners specify none at all or only those that may be required by law. Those required by law are imposed via a jurisdiction's building code. Almost all U.S. jurisdictions base their building codes on "model codes" developed by associations of building officials. The International Code Council (ICC) is the most prominent of these groups and its International Building Code (IBC) is the most commonly used model. As a result, many jurisdictions now require IBC "Special Inspection," a term defined by the IBC as "the required examination of the materials, installation, fabrication, erection, or placement of components and connections requiring special expertise to ensure compliance with approved construction documents and referenced standards." Special Inspection requirements vary from jurisdiction to jurisdiction based on the provisions adopted by the local building official. While some of the services involved may be similar to or the same as conventional CoMET services, Special Inspection is handled differently. Most commonly, the owner or the owner's agent is required to retain a building-official-approved Special Inspection-services provider. Special Inspection is often required to obtain a certificate of occupancy.

QC services are those applied by or on behalf of a constructor to ensure the constructor has attained conditions the constructor has contractually agreed to attain. Most CoMET consultants are engaged far more to provide QA services than QC services.

Many CoMET procedures are specified in standards developed by standards-developing organisations (SDOs) such as the American Society of Civil Engineers (ASCE), ASTM International, and American Concrete Institute (ACI), using standards-development protocols approved by the American National Standards Institute (ANSI) and/or the International Organisation for Standardisation (ISO). All such standards identify what is minimally required to conform. Likewise, several organisations have developed programs to accredit CoMET field and laboratory services to perform certain types of testing and inspection. Some of these programs are more comprehensive than others; e.g., requiring regular calibration of equipment, participation in proficiency testing programs, and implementation and documentation of a (quality) management system to demonstrate technical competence. As with all such programs, of course, accreditation identifies what is least acceptable. Many CoMET laboratories go far beyond minimum requirements in an effort to attain higher levels of quality.

A variety of organisations – including local building departments – have developed personnel-certification protocols and requirements. In many jurisdictions, only appropriately certified individuals are permitted to perform certain evaluations. Individuals typically are required to meet certain prerequisites for certification and must pass examinations, in some cases involving performance observation in the field. The prerequisite for higher degrees of certification often include a requirement that the individual has met requirements for a lower degree of certification (e.g., Soils Technician I is in some cases a prerequisite for Soils Technician II). It should be noted that field representatives are sometimes referred to as "soil testers," "technicians," "technicians/technologists," or "engineering technicians." ASFE/The Geoprofessional Business Association developed the term "field representative" to encompass all the many types of paraprofessionals involved (e.g., those involved with specific types of materials, such as reinforced concrete, soil, or steel; those who observe or inspect processes or conditions, such as welding inspectors, caisson inspectors, and foundation inspectors), and especially to underscore their significant, mutual responsibility, that purpose titles such as "technician" fail to signify. In fact, the engineers who direct CoMET operations are personally and professionally responsible and liable for their field representatives' acts and statements while representing the engineer on site.

Especially because CoMET consultants have more hands-on experience with construction activities than many other design-team members, many owners involve them (among other geoprofessionals) from the outset of a project, during the design phase, to help the owner and/or design team members develop technical specifications and establish testing and inspection requirements, instrumentation requirements and procedures, and observation programs. Geotechnical engineers employ CoMET services during the earliest stages of a project, to oversee subsurface sampling procedures, such as drilling.

Many of the CoMET services performed for construction projects are performed for environmental projects as well, but requirements tend to be less rigid because they involve fewer licensing and related requirements. For example, individuals may perform federally mandated all-appropriate inquiries – typically a phase-one environmental site assessment – without a license of any kind.

Other Geoprofessional Services

To the extent that archeology and paleontology require systematic subsurface excavation to recover artifacts, they, too, are considered geoprofessions. Many geoprofessional-services firms offer these services to those of their clients that need to satisfy federal and/or state regulations that require paleontological and/or archeological inquiry before site development or redevelopment activities can proceed.

Coastal Management

In some jurisdictions the terms sea defence and coastal protection are used to mean, respectively, defence against flooding and erosion. The term *coastal defence* is the more traditional term, but *coastal management* has become more popular as the field has expanded to include techniques that allow erosion to claim land.

Historical Background

Coastal engineering, as it relates to harbours, starts with the development of ancient civilizations together with the origin of maritime traffic, perhaps before 3500 B.C. Docks, breakwaters, and other harbour works were built by hand and often in a grand scale.

Some of the harbour works are still visible in a few of the harbours that exist today, while others have recently been explored by underwater archaeologists. Most of the grander ancient harbor works have disappeared following the fall of the Roman Empire.

Most ancient coastal efforts were directed to port structures, with the exception of a few places where life depended on coastline protection. Venice and its lagoon is one such case. Protection of the shore in Italy, England and the Netherlands can be traced back at least to the 6th century. The ancients understood such phenomena as the Mediterranean currents and wind patterns and the wind-wave cause-effect link.

The Romans introduced many revolutionary innovations in harbor design. They learned to build walls underwater and managed to construct solid breakwaters to protect fully exposed harbors. In some cases wave reflection may have been used to prevent silting. They also used low, water-surface breakwaters to trip the waves before they reached the main breakwater. They became the first dredgers in the Netherlands to maintain the harbour at Velsen. Silting problems here were solved when the previously sealed solid piers were replaced with new "open"-piled jetties. The Romans also introduced to the world the concept of the holiday at the coast.

Middle Age

The threat of attack from the sea caused many coastal towns and their harbours to be abandoned. Other harbours were lost due to natural causes such as rapid silting, shoreline advance or retreat, etc. The Venetian Lagoon was one of the few populated coastal areas with continuous prosperity and development where written reports document the evolution of coastal protection works. Engineering and scientific skills remained alive in the east, in Byzantium, where the Eastern Roman Empire survived for six hundred years while Western Rome decayed.

Modern Age

Erchin fundu could be considered the precursor of coastal engineering science, offering ideas and solutions often more than three centuries ahead of their common acceptance. Although great strides were made in the general scientific arena, little improvement was done beyond the Roman approach to harbour construction after the Renaissance. In the early 19th century, the advent of the steam engine, the search for new lands and trade routes, the expansion of the British Empire through her colonies, and other influences, all contributed to the revitalization of sea trade and a renewed interest in port works.

Twentieth Century

Evolution of shore protection and the shift from structures to beach nourishment. Prior to the 1950s, the general practice was to use hard structures to protect against beach erosion or storm damages. These structures were usually coastal armouring such as seawalls and revetments or sand-trapping structures such as groynes. During the 1920s and '30s, private or local community interests protected many areas of the shore using these techniques in a rather ad hoc manner. In certain resort areas, structures had proliferated to such an extent that the protection actually impeded the recreational use of the beaches. Erosion of the sand continued, but the fixed back-beach line remained, resulting in a loss of beach area. The obtrusiveness and cost of these structures led in the late 1940s and early 1950s, to move toward a new, more dynamic, method. Projects no longer relied solely on hard coastal defence structures, as techniques were developed which replicated the protective characteristics of natural beach and dune systems. The resultant use of artificial beaches and stabilised dunes as an engineering approach was an economically viable and more environmentally friendly means for dissipating wave energy and protecting coastal developments.

Over the past hundred years the limited knowledge of coastal sediment transport processes at the local authorities level has often resulted in inappropriate measures of coastal erosion mitigation. In many cases, measures may have solved coastal erosion locally but have exacerbated coastal erosion problems at other locations -up to tens of kilometres away- or have generated other environmental problems.

Current Challenges in Coastal Management

The coastal zone is a dynamic area of natural change and of increasing human use. They occupy less than 15% of the Earth's land surface; yet accommodate more than 50% of the world population (it is estimated that 3.1 billion people live within 200 kilometres from the sea). With three-quarters of the world population expected to reside in the coastal zone by 2025, human activities originating from this small land area will impose an inordinate amount of pressures on the global system. Coastal zones contain rich resources to produce goods and services and are home to most commercial and industrial activities. In the European Union, almost half of the population now lives within 50 kilometres of the sea and coastal zone resources produce much of the Union's economic wealth. The fishing, shipping and tourism industries all compete for vital space along Europe's estimated 89 000 kilometres of coastline, and coastal zones contain some of Europe's most fragile and valuable natural habitats. Shore protection consists up to the 50's of interposing a static structure between the sea and the land to prevent erosion and or flooding, and it has a long history. From that period new technical or friendly policies have been developed to preserve the environment when possible. Is already important where there are extensive low-lying areas that require protection. For instance: Venice, New Orleans, Nagara river in Japan, Holland, Caspian Sea Protection against the sea level rise in the 21st century will be especially important, as sea level rise is currently accelerating. This will be a challenge to coastal management, since seawalls and breakwaters are generally expensive to construct, and the costs to build protection in the face of sea-level rise would be enormous.

Changes on sea level have a direct adaptative response from beaches and coastal systems, as we can see in the succession of a lowering sea level. When the sea level rises, coastal sediments are in part pushed up by wave and tide energy, so sea-level rise processes have a component of sediment transport landwards. This results in a dynamic model of rise effects with a continuous sediment displacement that is not compatible with static models where coastline change is only based on topographic data.

Planning Approaches

There are five generic strategies for coastal defence:

- inaction leading to eventual abandonment
- Managed retreat or realignment, which plans for retreat and adopts engineering solutions that recognise natural processes of adjustment, and identifies a new line of defence where to construct new defences
- Hold the line, shoreline protection, whereby seawalls are constructed around the coastlines
- Move seawards, this happens by constructing new defences seaward the original ones
- Limited intervention, accommodation, by which adjustments are made to be able to cope with inundation, raising coastal land and buildings vertically.

The decision to choose a strategy is site-specific, depending on pattern of relative sea-level change, geomorphological setting, sediment availability and erosion, as well a series of social, economic and political factors.

Alternatively, integrated coastal zone management approaches may be used to prevent development in erosion- or flood-prone areas to begin with. Growth management can be a challenge for coastal local authorities who often struggle to provide the infrastructure required by new residents seeking seachange lifestyles. Sustainable transport investment to reduce the average footprint of coastal visitors is often a good way out of coastal gridlock. Examples include Dongtan and the Gold Coast Oceanway.

Do Nothing

The 'do nothing' option, involving no protection, is a cheap and expedient way to let the coast take care of itself. It involves the abandonment of coastal facilities when they are subject to coastal erosion, and either gradually landward retreat or evacuation and resettlement elsewhere. This option is very environmental friendly and the only pollution produced is from the resettlement process. However it does mean losing a lot of land to the sea and people will lose their homes.

Managed Retreat

Managed retreat is an alternative to constructing or maintaining coastal structures. Managed retreat allows an area that was not previously exposed to flooding by the sea to become flooded. This process is usually in low lying estuarine or deltaic areas and almost always involves flooding of land that has at some point in the past been reclaimed from the sea. Managed retreat is often a response to a change in sediment budget or to sea level rise. The technique is used when the land adjacent to the sea is low in value. A decision is made to allow the land to erode and flood, creating new sea, inter-tidal and salt-marsh habitats. This process may continue over many years and natural stabilisation will occur.

The earliest managed retreat in the UK was an area of 0.8 ha at Northey Island in Essex, that was flooded in 1991. This was followed by Tollesbury and Orplands in Essex, where the sea walls were breached in 1995. In the Ebro delta (Spain) coastal authorities have planned a managed retreat in response to coastal erosion (MMA 2005, Sitges, Meeting on Coastal Engineering; EUROSION project).

Cost – The main cost is generally the purchase of land to be flooded. Housings compensation for relocation of residents may be needed. Any other human made structure which will be engulfed by the sea may need to be safely dismantled to prevent sea pollution. In some cases, a retaining wall or bund must be constructed inland in order to protect land beyond the area to be flooded, although such structures can generally be lower than would be needed on the existing coast. Monitoring of the evolution of the flooded area is another cost. Costs may be lowest if existing defences are left to fail naturally, but often the realignment project will be more actively managed, for example by creating an artificial breach in existing defences to allow the sea in at a particular place in a controlled fashion, or by pre-forming drainage channels for created salt-marsh.

Hold the Line

Human strategies on the coast have been heavily based on a static engineered response, whereas the coast is in, or strives towards, a dynamic equilibrium (Schembri, 2009). Solid coastal structures are built and persist because they protect expensive properties or infrastructures, but they often relocate the problem downdrift or to another part of the coast. Soft options like beach nourishment, while also being temporary and needing regular replenishment, appear more acceptable, and go some way to restore the natural dynamism of the shoreline. However in many cases there is a legacy of decisions that were made in the past which have given rise to the present threats to coastal infrastructure and which necessitate immediate shore protection. For instance, the seawall and promenade of many coastal cities in Europe represents a highly engineered use of prime seafront flange-eating space, which might be preferably designated as public open space, parkland and amenities if it were available today. Such open space might also allow greater flexibility in terms of future land-use change, for instance through managed retreat, in the face of threats of erosion or inundation as a result of sea-level rise. Foredunes areas represent a natural reserve which can be called upon in the face of extreme events; building on these areas leaves little option but to undertake costly protective measures when extreme events (whether amplified by gradual global change or not) threaten. Managed retreat can comprise 'setbacks', rolling easements and other planning tools including building within a particular design life. Maintenance of those structures or soft techniques can arrive at a critical point (economically or environmental) to change adopted strategy.

- Structural or hard engineering techniques, i.e. using permanent concrete and rock constructions to "fix" the coastline and protect the assets locate behind. These techniques—seawalls, groynes, detached breakwaters, and revetments—represent a significant share of protected shoreline in Europe (more than 70%).
- Soft engineering techniques (e.g. sand nourishments), building with natural processes and relying on natural elements such as sands, dunes and vegetation to prevent erosive forces from reaching the backshore. These techniques include beach nourishment and sand dune stabilisation.

Move Seaward

The futility of trying to predict future scenarios where there is a large human influence is apparent. Even future climate is to a certain extent a function of what humans choose to make of it, for example by restricting greenhouse gas emissions to control climate change. In some cases - where new areas are needed for new economic or ecological development - a move seaward strategy can be adopted. Some examples from erosion are: Koge Bay (Dk) Western Scheldt estuary (NI), Chatelaillon (F), Ebro delta (E).

There is an obvious downside to this strategy. Coastal erosion is already widespread, and there are many coasts where exceptional high tides or storm surges result in encroachment on the shore, impinging on human activity. If the sea rises, many coasts that are developed with infrastructure along or close to the shoreline will be unable to accommodate erosion, and will experiment a so-called "coastal squeeze". This occurs where the ecological or geomorphological zones that would normally retreat landwards encounter solid structures and are squeezed out. Wetlands, salt marshes, mangroves and adjacent fresh water wetlands are particularly likely to suffer from this squeeze.

An upside to the strategy is that moving seaward (and upward) can create land of high value which can bring the investment required to cope with climate change.

Limited Intervention

Limited intervention is an action taken whereby the management only solves the problem to some extent, usually in areas of low economic significance. Measures taken using limited intervention often encourage the succession of haloseres, including salt marshes and sand dunes. This will normally result in the land behind the halosere being more sufficiently protected, as wave energy will be dissipated by the accumulated sediment and additional vegetation residing in the newly formed habitat. Although the new halosere is not strictly man-made, as many natural processes will contribute to the succession of the halosere, anthropogenic factors are partially responsible for the formation as an initial factor was needed to help start the process of succession. This must not be confused with 'accommodate' which is about property e.g. effective insurance, early warning systems and not about habitat.

Construction Techniques

The following is a catalogue of relevant techniques that could be employed as coastal management techniques. *The costs given are very rough estimates made during 2005, based on UK Pound sterling.*

Hard Engineering Methods

Groynes: Groynes are wooden often made of greenhart, concrete and/or rock barriers or walls perpendicular to the sea. Beach material builds up on the downdrift side, where lflittoral drift is predominantly in one direction, creating a wider and a more plentiful beach, therefore enhancing the protection for the coast because the sand material filters and absorbs the wave energy. However, there is a corresponding loss of beach material on the updrift side, requiring that another groyne to be built there. Moreover, groynes do not protect the beach against storm-driven waves and if placed too close together will create currents, which will carry sand material offshore.

Groynes are extremely cost-effective coastal defence measures, requiring little maintenance, and are one of the most common coastal defence structures. However, groynes are increasingly viewed as detrimental to the aesthetics of the coastline, and face strong opposition in many coastal communities. Many experts consider groynes to be a "soft" solution to coastal erosion because of the enhancement of the existing beach.

But groyne construction creates a problem known as Terminal Groyne Syndrome. The terminal groyne prevents longshore drift from bringing material to other nearby places. This is a common problem along the Hampshire and Sussex coastline in the UK; a perfect example is Worthing.

Sea Walls: Walls of concrete or rock, built at the base of a cliff or at the back of a beach, or used to protect a settlement against erosion or flooding. They are usually about 3-5 metres high. Older style vertical seawalls reflected all the energy of the waves back out to sea, and for this purpose were often given recurved crest walls which also increase the local turbulence, and thus increasing entrainment of sand and sediment. During storms, sea walls help longshore drift

Modern seawalls aim to re-direct most of the incident energy, resulting in low reflected waves and much reduced turbulence and thus take the form of sloping revetments. Current designs use porous designs of rock, concrete armour (Seabees, SHEDs, Xblocs) with intermediate flights of steps for beach access, whilst in places where high rates of pedestrian access are required, the steps take over the whole of the frontage, but at a flatter slope if the same crest levels are to be achieved.

Care needs to be taken in the location of a seawall, particularly in relation to the swept prism of the beach profile, the consequences of long term beach recession and amenity crest level. These factors must be considered in assessing the cost benefit ratio, which must be favourable in order to justify construction of a seawall.

Sea walls can cause beaches to dissipate rendering them useless for beach goers. Their presence also scars the very landscape that they are trying to save.

Modern examples can be found at Cronulla (NSW, 1985-6), Blackpool (1986–2001), Lincolnshire (1992–1997) & Wallasey (1983–1993). The sites at Blackpool and Cronulla can be visited both by Google Earth and by local webcams (Cronulla, Cleveleys).

A most interesting example is the seawall at Sandwich, Kent, where the Seabee seawall is buried at the back of the beach under the shingle with crest level at road kerb level.

Sea walls are probably the second most traditional method used in coastal management.

Sea walls cost £10,000 per metre (depending on material, height and width)

Revetments

Wooden slanted or upright blockades, built parallel to the sea on the coast, usually towards the back of the beach to protect the cliff or settlement beyond. The most basic revetments consist of timber slants with a possible rock infill. Waves break against the revetments, which dissipate and absorb the energy. The cliff base is protected by the beach material held behind the barriers, as the revetments trap some of the material. They may be watertight, covering the slope completely, or porous, to allow water to filter through after the wave energy has been dissipated. Most revetments do not significantly interfere with transport of longshore drift. Since the wall greatly absorbs the energy instead of reflecting, it erodes and destroys the revetment structure; therefore, major maintenance will be needed within a moderate time of being built, this will be greatly determined by the material the structure was built with and the quality of the product.

The *Cost* – Confirmed by material used; est. \$2340 – \$4000. Average \$10 per metre built.

Rock Armour

Also known as riprap, rock armour is large rocks piled or placed at the foot of dunes or cliffs with native stones of the beach. This is generally used in areas prone to erosion to absorb the wave energy and hold beach material. Although effective, this solution is unpopular due to the fact that it is unsightly. Also, longshore drift is not hindered. Rock armour has a limited lifespan, it is not effective in storm conditions, and it reduces the recreational value of a beach. The cost is around £3000 per metre, depending on the type of rocks used.

Gabions

Boulders and rocks are wired into mesh cages and usually placed in front of areas vulnerable to heavy erosion: sometimes at cliffs edges or jag out at a right angle to the beach like a large groyne. When the seawater breaks on the gabion, the water drains through leaving sediments, also the rocks and boulders absorb a moderate amount of the wave energy. Gabions need to be securely tied to prevent abrasion of wire by rocks, or detachment of plastic coating by stretching. Hexagonal mesh distributes overloads better than rectangular mesh.

Downsides include wear rates and visually intrusiveness.

Cost – est. £11 per m.

Offshore Breakwater

Enormous concrete blocks and natural boulders are sunk offshore to alter wave direction and to filter the energy of waves and tides. The waves break further offshore and therefore reduce their erosive power. This leads to wider beaches, which absorb the reduced wave energy, protecting cliff and settlements behind. The Dolos which was invented by a South African engineer in East London has replaced the use of enormous concrete blocks because the dolos is much more resistant to wave action and requires less concrete to produce a superior result. Similar concrete objects like the Dolos are the A-jack, Akmon, Xbloc and the Tetrapod, Accropode.

Cost – est. £1,950 per m. Water depth may increase the cost.

Cliff Stabilisation

Cliff stabilisation can be accomplished through drainage of excess rainwater of through terracing, planting, and wiring to hold cliffs in place. Cliff drainage is used to hold a cliff together using plants, fences and terracing, this is used to help prevent landslides and other natural disasters.

Entrance Training Walls

Rock or concrete walls built to constrain a river or creek discharging across a sandy coastline. The walls help to stabilise and deepen the channel which benefits navigation, flood management, river erosion and water quality but can cause coastal erosion due to the interruption of longshore drift. One solution is the installation of a sand bypassing system to pump sand under and around the entrance training walls.

Cost – Expensive - Gold Coast Seaway was a A$50M project in the 1980s and the adjacent sand bypassing project costs A$3M per year to pump 500,000 cubic metres of sand across the trained entrance.

Floodgates

Storm surge barriers, or floodgates, were introduced after the North Sea Flood of 1953 and are a prophylactic method to prevent damage from storm surges. They are habitually open and allow free passage, but close when the land is under threat of a storm surge. The Thames Barrier is an example of such a structure.

Soft Engineering Methods

Beach Replenishment: Beach replenishment or nourishment is one of the most popular soft engineering techniques of coastal defence management schemes. This involves importing sand off the beach and piling it on top of the existing sand. The imported sand must be of a similar quality to the existing beach material so it can integrate with the natural processes occurring there, without causing any adverse effects. Beach nourishment can be used alongside the groyne schemes. The scheme requires constant maintenance: 1 to 10 year life before first major recharge. *Cost* – est. £5,000-£200,000 per 100 metre, plus control structures, ongoing management and minor works.

Sand Dune Stabilisation

Vegetation can be used to encourage dune growth by trapping and stabilising blown sand.

Cost – est. of £1.1 million per annum.

Beach Drainage

Beach drainage or beach face dewatering lowers the water table locally beneath the beach face. This causes accretion of sand above the drainage system.

Grant (1946) – the elevation of the beach watertable had an important bearing on deposition and erosion across the foreshore. A high watertable coincided with periods of accelerated beach erosion, and conversely, a low watertable coincided with pronounced aggradation of the foreshore A lower watertable (unsaturated beach face) facilitates deposition by reducing flow velocities during backwash and prolonging laminar flow. In contrast, a high watertable results in condition favouring beach erosion. With the beach in a saturated state, Grant proposed that backwash velocity is accelerated by the addition of groundwater seepage out of the beach within the effluent zone.

Turner and Leatherman (1997) moving from the origins and development of the dewatering concept to field and laboratory studies available at the time of writing concluded that there was too little evidence for being convinced that the systems had a positive effect. None of the case studies provide full scientific evidence of indisputable positive results regarding beach stabilisation although in some cases an overall positive performance was reported. In many cases no adequate long-term monitoring was undertaken at a frequency high enough to discriminate the response to high energy erosive events.

A useful side effect of the system is that the collected seawater is very pure because of the sand filtration effect. It may be discharged back to sea but can also be used to oxygenate stagnant inland lagoons /marinas or used as feed for heat pumps, desalination plants, land-based aquaculture, aquariums or seawater swimming pools.

Beach drainage systems have been installed in many locations around the world to halt and reverse erosion trends in sand beaches. Twenty four beach drainage systems have been installed since 1981 in Denmark, USA, UK, Japan, Spain, Sweden, France, Italy and Malaysia.

Costs

The costs of installation and operation per metre of shoreline protection will vary due to

- system length (non-linear cost elements)
- pump flow rates (sand permeability, power costs)
- soil conditions (presence of rock or impermeable strata)
- discharge arrangement /filtered seawater utilisation
- drainage design, materials selection & installation methods
- geographical considerations (location logistics)
- regional economic considerations (local capabilities /costs)
- study requirements /consent process.

The costs associated with a beach drainage system are generally considerably lower than hard engineered structures. They also compare very favourably with beach nourishment projects, particularly when long-term project economics are considered (nourishment projects often have a limited life or a program of re-nourishment).

Monitoring Coastal Zones

Coastal zone managers are faced with difficult and complex choices about how best to reduce property damage in the shorelines. One of the problems they face is error and uncertainty in the information available to them on the processes that cause erosion of beaches. Video-based monitoring lets collect data continuously at low cost and produce analyses of shoreline processes over a wide range of averaging intervals.

Event Warning Systems

Event warning systems, such as tsunami warnings and storm surge warnings, can be used to minimise the human impact of catastrophic events that cause coastal erosion. Storm surge warnings can also be used to determine when to close floodgates to reduce the physical impact of such events.

Wireless sensor networks can be deployed quickly to set up a coastal erosion monitoring system, and scaled accordingly.

Shoreline Mapping

Defining the shoreline is a difficult task due to the dynamic nature of the coast and the intended application of the shoreline. Given this idea the shoreline must therefore be considered in a temporal sense whereby the scale is dependent on the context of the investigation (Boak & Turner 2005).

The following definition of the coast and shoreline is most commonly employed for the purposes of shoreline mapping. The coast comprises the interface between land and sea, and the shoreline is represented by the margin between the two (Woodroffe, 2002). Due to the dynamic nature of the shoreline coastal investigators adopt the use of shoreline indicators to represent the true shoreline position (Boak & Turner 2005).

Shoreline Indicator

The choice of shoreline indictor is a primary consideration in shoreline mapping. According to Leatherman (2003) it is important that indicators are easily identified in the field and on aerial photography. Shoreline indicators may be physical beach morphological features such as the berm crest, scarp edge, vegetation line, dune toe, dune crest and cliff or the bluff crest and toe. Alternatively, non-morphological features may also be used. These indicators are based on water level including the high water line, mean high water line, wet/dry boundary, and the physical water line.

The high water line (HWL), defined as the wet/dry line is the most commonly used shoreline indicator because it is visible in the field, and can be interpreted on both colour and grey scale aerial photographs (Leatherman, 2003; Crowell et al. 1991). The HWL represents the landward extent of the most recent high tide and is characterised by a change in sand colour due to repeated, periodic inundation by high tides. The HWL is portrayed on aerial photographs by the most landward change in colour or grey tone.

Importance and Application

The location of the shoreline and its changing position over time is of fundamental importance to coastal scientists, engineers and managers. Present day shoreline monitoring campaigns provide information about historic shoreline location and movement, and about predictions of future change (Appeaning Addo et al. 2008).

More specifically the position of the shoreline in the past, at present and where it is predicted to be in the future is useful for in the design of coastal protection, to calibrate and verify numerical models to assess sea level rise, map hazard zones and formulate policies to regulate coastal development. Accurate and consistent delineation of the shoreline is integral to all of these tasks. The location of the shoreline also provides information regarding shoreline reorientation adjacent to structures, beachwidth, volume and rates of historical change.

Data Sources

A variety of data sources are available for examining shoreline position however, the availability of historical data is limited at many coastal sites and so the choice of data source is largely limited to what is available for the site at a given time (Boak & Turner 2005). Shoreline mapping techniques applied to data sources have moved towards automation in association with technological advances and the need to reduce uncertainty. Although these changes have resulted in improvement in coastal data processing and storage capabilities, the frequent change in technology has prevented the emergence of one standard method of shoreline mapping. This has occurred because each data source and associated method have their own unique capabilities and shortcomings (Moore 2000). A number of the data sources used for shoreline mapping and their associated advantages and disadvantages are discussed below.

Historical Maps

In the event that a study requires the shoreline position to be mapped before the development of aerial photographs, or if the location has poor photograph coverage it is necessary to employ historical maps in order to detail shoreline position (Moore 2000). The main advantage and reason for using historical maps is that they are able to provide a historic record that is not available from other data sources. Many potential errors however are associated with historical coastal maps and charts. Such errors may be associated with scale, datum changes, distortions from uneven shrinkage, stretching, creases, tears and folds, different surveying standards, different publication standards, and projection errors (Boak & Turner 2005). The severity of these errors depends on the accuracy standards met by each map and the physical changes that have occurred since the publication of the map (Anders & Byrnes 1991). The oldest reliable source of shoreline data in the United States dates back to the early-to-mid-19th century and is the U.S Coast and Geodetic Survey/National Ocean Service T-sheets (Morton 1991). In the United Kingdom, many maps and charts were deemed to be inaccurate until around 1750. The founding of the Ordnance Survey in 1971 has since improved the accuracy of the mapping.

Aerial Photographs

Aerial photographs have been used since the 1920s to provide topographical information about an area. They are therefore a good database for compilation of historical shoreline change maps. Aerial photographs are the most commonly used data source in shoreline mapping because many coastal areas have extensive aerial photo coverage therefore providing a valuable record of shoreline position (Moore 2000).

In general, aerial photographs provide good spatial coverage of the coast however temporal coverage is very much site specific depending on the flight path of the aeroplane. A second disadvantage associated with aerial photography is that the interpretation of the shoreline position is subjective given the dynamic nature of the coastal environment. This combined with various distortions inherent in aerial photographs can lead to significant error levels (Moore 2000). The minimisation of further errors is discussed below.

Object Space Displacements

Conditions outside of the camera can cause objects in an image to be displaced from their true ground position. Such conditions may include ground relief, camera tilt and atmospheric refraction.

Relief displacement is prominent when photographing a variety of elevations. This situation causes objects above ground level to be displaced outward from the centre of the photograph and objects below ground level to be displaced toward the centre of the image. The severity of the displacement is affected negatively with decreases in flight altitude and as radial distance from the centre of the photograph increases.

This distortion can be minimised by photographing numerous swaths and creating a mosaic of the images. This technique will create a focus for the centre of each photograph where distortion is minimised. It is important to note that this error is not common in shoreline mapping is the relief is fairly constant. It is however important to consider when mapping cliffs (Moore 2000).

Ideally aerial photographs are taken so the optical axis of the camera is perfectly perpendicular to the ground surface thereby creating a vertical photograph. Unfortunately this is not often the case and virtually all aerial photographs experience tilt whereby up to 3° is not uncommon (Camfield et al. 1996). In this situation the scale of the image will be larger on the upward side of the tilt axis and smaller on the downward side. Moore, (2000) notes that many coastal researchers have not realised the severity of this error and therefore do not consider it in their methods.

Radial Lens Distortion

Lens distortion varies as a function of radial distance from the iso-centre of the photograph meaning that the centre of the image is relatively distortion free, but as the angle of view increases the distortion becomes more prominent. This is a significant source of error in earlier aerial photography but as technology has increased and camera lens have become more refined it has become less of an issue with later photographs. Such a distortion is impossible to correct for without knowing the make and model of the lens used to capture the image. However if overlapping images have been acquired one can digitize the centre portions of the aerial photographs (Crowell et al. 1991).

Delineation of the Shoreline

The dynamic nature of the coast has meant that accurate mapping of an instantaneous shoreline position has been associated with significant uncertainty. This uncertainty arises because at any given time the position of the shoreline is influenced by the short-term effect of the tide and a wide variety of long term effects such as relative sea-level rise and alongshore littoral sediment movement. Not only does this affect the accuracy of computed historic shoreline position but also any predicted future positions (Appeaning Addo et al. 2008). As mentioned earlier the HWL is most commonly used as a shoreline indicator.

This can usually be seen as a significant tonal change on aerial photographs. There are however many errors associated with using the wet/dry line as a proxy for the HWL and shoreline. The errors of largest concern are the short term migration of the wet/dry line, interpretation of the wet/dry line on a photograph and measurement of the interpreted line position. Systematic errors such as the migration of the wet/dry line may arise from tidal and seasonal changes. Storm-induced erosion is another factor which may cause the wet/dry line to migrate landward. Field investigations have shown that these changes can be minimised by using only summertime data. Furthermore, the error bar can be significantly reduced by using the longest record of reliable data to calculate erosion rates (Leatherman 2003). Finally it is important to note that errors may arise due to the difficulty of measuring a single line on a photograph. For example where the pen line is 0.13 mm thick this translates to an error of ±2.6 m on a 1:20000 scale photograph.

Beach Profiling Surveys

Beach profiling surveys are typically repeated at regular intervals along the coast in order to measure short-term (daily to annual) variations in shoreline position and beach volume. (Smith & Zarillo 1990). Beach profiling is a very accurate source of information however measurements are generally subject to the limitations of conventional surveying techniques. Shoreline data derived from beach profiling is often spatially and temporally limited due to the high cost associated with such a labour intensive activity. Shorelines are generally derived by interpolating between a series of discrete beach profiles. It is important to note however that the distance between the profiles is usually quite large and so the accuracy of the interpolating becomes compromised. In contrast to aerial photographs, survey data is limited to smaller lengths of shoreline generally less than ten kilometres (Boak & Turner 2005). Beach profiling data is commonly available in from regional councils in New Zealand such as those compiled by the Hawkes Bay Regional Council.

Remote Sensing

Technological advancement over the last decade has led to the development of a range of airborne, satellite and land based remote sensing techniques (Smith & Zarillo 1990). Some of the remotely sensed data sources are listed below:

- Multispectral and hyperspectral imaging
- Microwave sensors
- Global positioning system (GPS)
- Airborne light detection and ranging technology (LIDAR).

Remote sensing techniques are attractive as they are cost effective, reduce manual error and remove the subjective approach of conventional field techniques (Maiti et al. 2009). Remote sensing is a relatively new concept and so extensive historical observations are unavailable. Given this idea, it is important that coastal morphology observations are quantified by coupling remotely sensed data with other sources of information detailing historic shoreline position from archived sources.

Video Analysis

Video analysis provides quantitative, cost-effective, continuous and long-term monitoring beaches (Turner et al. 2004). The advancement of coastal video systems over the past 15 years has resulted in the extraction of large amounts of geophysical data from images. Such data includes that about coastal morphology, surface currents and wave parameters. The main advantage of video analysis lies in the ability to reliably quantify these parameters with high resolution and coverage in both space and time. This in particular highlights their potential importance as an effective coastal monitoring system and an aid to coastal zone management (Van Koningsveld et al. 2007). Interesting case studies have been carried out using video analysis. Turner et al. (2004) used a video-based ARGUS coastal imaging system to monitor and quantify the regional-scale coastal response to sand nourishment and construction of the world-first Gold Coastartificial (surfing) reef in Australia. In addition, Smit et al. (2007) demonstrated the added value of high resolution video observations for making short-term predictions of near shore hydrodynamic and morphological processes, at temporal scales of metres to kilometres and days to seasons.

Geotechnical Engineering

Geotechnical engineering is the branch of civil engineering concerned with the engineering behaviour of earth materials. Geotechnical engineering is important in civil engineering, but is also used by military, mining, petroleum, or any other engineering concerned with construction on or in the ground. Geotechnical engineering uses principles of soil mechanics and rock mechanics to investigate subsurface conditions and materials; determine the relevant physical/ mechanical and chemical properties of these materials; evaluate stability of natural slopes and man-made soil deposits; assess risks posed by site conditions; design earthworks and structure foundations; and monitor site conditions, earthwork and foundation construction.

A typical geotechnical engineering project begins with a review of project needs to define the required material properties. Then follows a site investigation of soil, rock, fault distribution and bedrock properties on and below an area of interest to determine their engineering properties including how they will interact with, on or in a proposed construction. Site investigations are needed to gain an understanding of the area in or on which the engineering will take place. Investigations can include the assessment of the risk to humans, property and the environment from natural hazards such as earthquakes, landslides, sinkholes, soil liquefaction, debris flows and rockfalls.

Ground Improvement refers to a technique that improves the engineering properties of the soil mass treated. Usually, the properties that are modified are shear strength, stiffness and permeability. Ground improvement has developed into a sophisticated tool to support foundations for a wide variety of structures. Properly applied, i.e. after giving due to consideration to the nature of the ground being improved and the type and sensitivity of the structures being built, ground improvement often reduces directs costs and saves time.

A geotechnical engineer then determines and designs the type of foundations, earthworks, and/or pavement subgrades required for the intended man-made structures to be built. Foundations are designed and constructed for structures of various sizes such as high-rise buildings, bridges, medium to large commercial buildings, and smaller structures where the soil conditions do not allow code-based design.

Foundations built for above-ground structures include shallow and deep foundations. Retaining structures include earth-filled dams and retaining walls. Earthworks include embankments, tunnels, dikes, levees, channels, reservoirs, deposition of hazardous waste and sanitary landfills.

Geotechnical engineering is also related to coastal and ocean engineering. Coastal engineering can involve the design and construction of wharves, marinas, and jetties. Ocean engineering can involve foundation and anchor systems for offshore structures such as oil platforms.

The fields of geotechnical engineering and engineering geology are closely related, and have large areas of overlap. However, the field of geotechnical engineering is a speciality of engineering, where the field of engineering geology is a speciality of geology.

History

Humans have historically used soil as a material for flood control, irrigation purposes, burial sites, building foundations, and as construction material for buildings. First activities were linked to irrigation and flood control, as demonstrated by traces of dykes, dams, and canals dating back to at least 2000 BCE that were found in ancient Egypt, ancient Mesopotamia and the Fertile Crescent, as well as around the early settlements of Mohenjo Daro and Harappa in the Indus valley. As the cities expanded, structures were erected supported by formalised foundations; Ancient Greeks notably constructed pad footings and strip-and-raft foundations. Until the 18th century, however, no theoretical basis for soil design had been developed and the discipline was more of an art than a science, relying on past experience.

Several foundation-related engineering problems, such as the Leaning Tower of Pisa, prompted scientists to begin taking a more scientific-based approach to examining the subsurface. The earliest advances occurred in the development of earth pressure theories for the construction of retaining walls. Henri Gautier, a French Royal Engineer, recognised the "natural slope" of different soils in 1717, an idea later known as the soil's angle of repose. A rudimentary soil classification system was also developed based on a material's unit weight, which is no longer considered a good indication of soil type.

The application of the principles of mechanics to soils was documented as early as 1773 when Charles Coulomb (a physicist, engineer, and army Captain) developed improved methods to determine the earth pressures against military ramparts. Coulomb observed that, at failure, a distinct slip plane would form behind a sliding retaining wall and he suggested that the maximum shear stress on the slip plane, for design purposes, was the sum of the soil cohesion, *c*, and friction , where is the normal stress on the slip plane and is the friction angle of the soil. By combining Coulomb's theory with Christian Otto Mohr's 2D stress state, the theory became known as Mohr-Coulomb theory. Although it is now recognised that precise determination of cohesion is impossible because *c* is not a fundamental soil property, the Mohr-Coulomb theory is still used in practice today.

In the 19th century Henry Darcy developed what is now known as Darcy's Law describing the flow of fluids in porous media. Joseph Boussinesq (a mathematician and physicist) developed theories of stress distribution in elastic solids that proved useful for estimating stresses at depth in the ground; William Rankine, an engineer and physicist, developed an alternative to Coulomb's earth pressure theory. Albert Atterberg developed the clay consistency indices that are still used today for soil classification. Osborne Reynolds recognised in 1885 that shearing causes volumetric dilation of dense and contraction of loose granular materials.

Modern geotechnical engineering is said to have begun in 1925 with the publication of *Erdbaumechanik* by Karl Terzaghi (a civil engineer and geologist). Considered by many to be the father of modern soil mechanics and geotechnical engineering, Terzaghi developed the principle of effective stress, and demonstrated that the shear strength of soil is controlled by effective stress. Terzaghi also developed the framework for theories of bearing capacity of foundations, and the theory for prediction of the rate of settlement of clay layers due to consolidation. In his 1948 book, Donald Taylor recognised that interlocking and dilation of densely packed particles contributed to the peak strength of a soil. The interrelationships between volume change behaviour (dilation, contraction, and consolidation) and shearing behaviour were all connected via the theory of plasticity using critical state soil mechanics by Roscoe, Schofield, and Wroth with the publication of "On the Yielding of Soils" in 1958. Critical state soil mechanics is the basis for many contemporary advanced constitutive models describing the behaviour of soil.

Practicing Engineers

Geotechnical engineers are typically graduates of a four-year civil engineering program and often hold a masters degree. In the USA, geotechnical engineers are typically licensed and regulated as Professional Engineers (PEs) in most states; currently only California and Oregon have licensed geotechnical engineering specialities. State governments will typically license engineers who have graduated from an ABET accredited school, passed the Fundamentals of Engineering examination, completed several years of work experience under the supervision of a licensed Professional Engineer, and passed the Professional Engineering examination.

Soil Mechanics

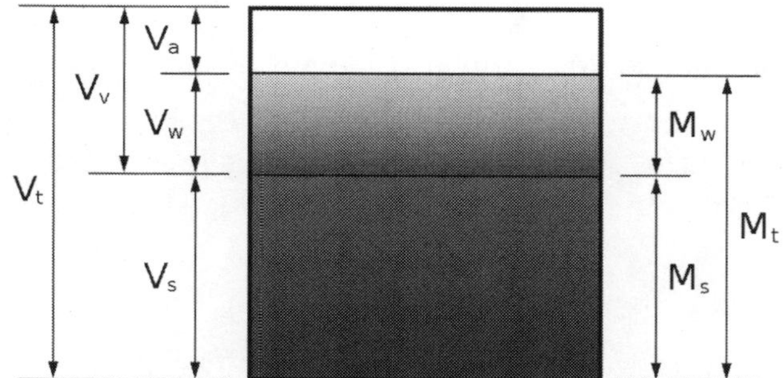

Figure: *A phase diagram of soil indicating the weights and volumes of air, soil, water, and voids.*

In geotechnical engineering, soils are considered a three-phase material composed of: rock or mineral particles, water and air. The voids of a soil, the spaces in between mineral particles, contain the water and air.

The engineering properties of soils are affected by four main factors: the predominant size of the mineral particles, the type of mineral particles, the grain size distribution, and the relative quantities of mineral, water and air present in the soil matrix. Fine particles (fines) are defined as particles less than 0.075 mm in diametre.

Soil Properties

Some of the important properties of soils that are used by geotechnical engineers to analyse site conditions and design earthworks, retaining structures, and foundations are:

Unit Weight

Total unit weight: Cumulative weight of the solid particles, water and air in the material per unit volume. Note that the air phase is often assumed to be weightless.

Porosity

Ratio of the volume of voids (containing air, water, or other fluids) in a soil to the total volume of the soil. A porosity of 0 implies that there are no voids in the soil.

Void Ratio

Void ratio is the ratio of the volume of voids to the volume of solid particles in a soil. Void ratio is mathematically related to the porosity.

Permeability

A measure of the ability of water to flow through the soil, expressed in units of velocity.

Compressibility

The rate of change of volume with effective stress. If the pores are filled with water, then the water must be squeezed out of the pores to allow volumetric compression of the soil; this process is called consolidation.

Shear Strength

The shear stress that will cause shear failure.

Atterberg Limits

Liquid limit, plastic limit, and shrinkage limit. These indices are used for estimation of other engineering properties and for soil classification.

Geotechnical Investigation

Geotechnical engineers perform geotechnical investigations to obtain information on the physical properties of soil and rock underlying (and sometimes adjacent to) a site to design earthworks and foundations for proposed structures, and for repair of distress to earthworks and structures caused by subsurface conditions. A geotechnical investigation will include surface exploration and subsurface exploration of a site. Sometimes, geophysical methods are used to obtain data about sites. Subsurface exploration usually involves in-situ testing (two common examples of in-situ tests are the standard penetration test and cone penetration test.

In addition site investigation will often include subsurface sampling and laboratory testing of the soil samples retrieved. The digging of test pits and trenching (particularly for locating faults and slide planes) may also be used to learn about soil conditions at depth. Large diametre borings are rarely used due to safety concerns and expense, but are sometimes used to allow a geologist or engineer to be lowered into the borehole for direct visual and manual examination of the soil and rock stratigraphy.

A variety of soil samplers exist to meet the needs of different engineering projects. The standard penetration test (SPT), which uses a thick-walled split spoon sampler, is the most common way to collect disturbed samples. Piston samplers, employing a thin-walled tube, are most commonly used for the collection of less disturbed samples. More advanced methods, such as ground freezing and the Sherbrooke block sampler, are superior, but even more expensive.

Atterberg limits tests, water content measurements, and grain size analysis, for example, may be performed on disturbed samples obtained from thick walled soil samplers. Properties such as shear strength, stiffness hydraulic conductivity, and coefficient of consolidation may be significantly altered by sample disturbance. To measure these properties in the laboratory, high quality sampling is required. Common tests to measure the strength and stiffness include the triaxial shear and unconfined compression test.

Surface exploration can include geologic mapping, geophysical methods, and photogrammetry; or it can be as simple as an engineer walking around to observe the physical conditions at the site. Geologic mapping and interpretation of geomorphology is typically completed in consultation with a geologist or engineering geologist.

Geophysical exploration is also sometimes used. Geophysical techniques used for subsurface exploration include measurement of seismic waves (pressure, shear, and Rayleigh waves), surface-wave methods and/or downhole methods, and electromagnetic surveys (magnetometre, resistivity, and ground-penetrating radar).

Foundations

A building's foundation transmits loads from buildings and other structures to the earth. Geotechnical engineers design foundations based on the load characteristics of the structure and the properties of the soils and/or bedrock at the site. In general, geotechnical engineers: 1) Estimate the magnitude and location of the loads to be supported;

2) Develop an investigation plan to explore the subsurface; 3) Determine necessary soil parameters through field and lab testing (e.g.,consolidation test, triaxial shear test, vane shear test, standard penetration test); 4) Design the foundation in the safest and most economical manner.

The primary considerations for foundation support are bearing capacity, settlement, and ground movement beneath the foundations. Bearing capacity is the ability of the site soils to support the loads imposed by buildings or structures. Settlement occurs under all foundations in all soil conditions, though lightly loaded structures or rock sites may experience negligible settlements. For heavier structures or softer sites, both overall settlement relative to unbuilt areas or neighbouring buildings, and differential settlement under a single structure, can be concerns. Of particular concern is settlement which occurs over time, as immediate settlement can usually be compensated for during construction. Ground movement beneath a structure's foundations can occur due to shrinkage or swell of expansive soils due to climatic changes, frost expansion of soil, melting of permafrost, slope instability, or other causes. All these factors must be considered during design of foundations.

Many building codes specify basic foundation design parameters for simple conditions, frequently varying by jurisdiction, but such design techniques are normally limited to certain types of construction and certain types of sites, and are frequently very conservative.

In areas of shallow bedrock, most foundations may bear directly on bedrock; in other areas, the soil may provide sufficient strength for the support of structures. In areas of deeper bedrock with soft overlying soils, deep foundations are used to support structures directly on the bedrock; in areas where bedrock is not economically available, stiff "bearing layers" are used to support deep foundations instead.

Shallow Foundations

Shallow foundations are a type of foundation that transfers building load to the very near the surface, rather than to a subsurface layer. Shallow foundations typically have a depth to width ratio of less than 1.

Footings

Footings (often called "spread footings" because they spread the load) are structural elements which transfer structure loads to the ground by direct a real contact. Footings can be isolated footings for point or column loads, or strip footings for wall or other long (line) loads. Footings are normally constructed from reinforced concrete cast directly onto the soil, and are typically embedded into the ground to penetrate through the zone of frost movement and/or to obtain additional bearing capacity.

Slab Foundations

A variant on spread footings is to have the entire structure bear on a single slab of concrete underlying the entire area of the structure. Slabs must be thick enough to provide sufficient rigidity to spread the bearing loads somewhat uniformly, and to minimise differential settlement across the foundation. In some cases, flexure is allowed and the building is constructed to tolerate small movements of the foundation instead. For small structures, like single-family houses, the slab may be less than 300 mm thick; for larger structures, the foundation slab may be several metres thick. Slab foundations can be either slab-on-grade foundations or embedded foundations, typically in buildings with basements. Slab-on-grade foundations must be designed to allow for potential ground movement due to changing soil conditions.

Deep Foundations

Deep foundations are used for structures or heavy loads when shallow foundations cannot provide adequate capacity, due to size and structural limitations. They may also be used to transfer building loads past weak or compressible soil layers. While shallow foundations rely solely on the bearing capacity of the soil beneath them, deep foundations can rely on end bearing resistance, frictional resistance along their length, or both in developing the required capacity. Geotechnical engineers use specialised tools, such as the cone penetration test, to estimate the amount of skin and end bearing resistance available in the subsurface.

There are many types of deep foundations including piles, drilled shafts, caissons, piers, and earth stabilised columns. Large buildings such as skyscrapers typically require deep foundations. For example, the Jin Mao Tower in China uses tubular steel piles about 1m (3.3 feet) driven to a depth of 83.5m (274 feet) to support its weight.

In buildings that are constructed and found to undergo settlement, underpinning piles can be used to stabilise the existing building.

Lateral Earth Support Structures

A retaining wall is a structure that holds back earth. Retaining walls stabilise soil and rock from downslope movement or erosion and provide support for vertical or near-vertical grade changes. Cofferdams and bulkheads, structures to hold back water, are sometimes also considered retaining walls.

The primary geotechnical concern in design and installation of retaining walls is that the retained material is attempting to move forward and downslope due to gravity. This creates soil pressure behind the wall, which can be analysed based on the angle of internal friction (φ) and the cohesive strength (c) of the material and the amount of allowable movement of the wall. This pressure is smallest at the top and increases toward the bottom in a manner similar to hydraulic pressure, and tends to push the wall forward and overturn it. Groundwater behind the wall that is not dissipated by a drainage system causes an additional horizontal hydraulic pressure on the wall.

Gravity Walls

Gravity walls depend on the size and weight of the wall mass to resist pressures from behind. Gravity walls will often have a slight setback, or batter, to improve wall stability. For short, landscaping walls, gravity walls made from dry-stacked (mortarless) stone or segmental concrete units (masonry units) are commonly used.

Earlier in the 20th century, taller retaining walls were often gravity walls made from large masses of concrete or stone. Today, taller retaining walls are increasingly built as composite gravity walls such as: geosynthetic or steel-reinforced backfill soil with precast facing; gabions (stacked steel wire baskets filled with rocks), crib walls (cells built up log cabin style from precast concrete or timber and filled with soil or free draining gravel) or soil-nailed walls (soil reinforced in place with steel and concrete rods).

For reinforced-soil gravity walls, the soil reinforcement is placed in horizontal layers throughout the height of the wall. Commonly, the soil reinforcement is geogrid, a high-strength polymer mesh, that provide tensile strength to hold soil together. The wall face is often of precast, segmental concrete units that can tolerate some differential movement. The reinforced soil's mass, along with the facing, becomes the gravity wall. The reinforced mass must be built large enough to retain the pressures from the soil behind it. Gravity walls usually must be a minimum of 30 to 40 percent as deep (thick) as the height of the wall, and may have to be larger if there is a slope or surcharge on the wall.

Cantilever Walls

Prior to the introduction of modern reinforced-soil gravity walls, cantilevered walls were the most common type of taller retaining wall.

Cantilevered walls are made from a relatively thin stem of steel-reinforced, cast-in-place concrete or mortared masonry (often in the shape of an inverted T). These walls cantilever loads (like a beam) to a large, structural footing; converting horizontal pressures from behind the wall to vertical pressures on the ground below. Sometimes cantilevered walls are buttressed on the front, or include a counterfort on the back, to improve their stability against high loads. Buttresses are short wing walls at right angles to the main trend of the wall. These walls require rigid concrete footings below seasonal frost depth. This type of wall uses much less material than a traditional gravity wall. Cantilever walls resist lateral pressures by friction at the base of the wall and/or passive earth pressure, the tendency of the soil to resist lateral movement. Basements are a form of cantilever walls, but the forces on the basement walls are greater than on conventional walls because the basement wall is not free to move.

Excavation Shoring

Shoring of temporary excavations frequently requires a wall design which does not extend laterally beyond the wall, so shoring extends below the planned base of the excavation. Common methods of shoring are the use of sheet piles or soldier beams and lagging. Sheet piles are a form of driven piling using thin interlocking sheets of steel to obtain a continuous barrier in the ground, and are driven prior to excavation. Soldier beams are constructed of wide flange steel H sections spaced about 2–3 m apart, driven prior to excavation. As the excavation proceeds, horizontal timber or steel sheeting (lagging) is inserted behind the H pile flanges.

In some cases, the lateral support which can be provided by the shoring wall alone is insufficient to resist the planned lateral loads; in this case additional support is provided by walers or tie-backs. Walers are structural elements which connect across the excavation so that the loads from the soil on either side of the excavation are used to resist each other, or which transfer horizontal loads from the shoring wall to the base of the excavation. Tie-backs are steel tendons drilled into the face of the wall which extend beyond the soil which is applying pressure to the wall, to provide additional lateral resistance to the wall.

Earth Structures

Compaction: Compaction is the process by which the strength and stiffness of soil may be increased and permeability may be decreased. Fill placement work often has specifications requiring a specific degree of compaction, or alternatively, specific properties of the compacted soil. In-situ soils can be compacted either by excavation and recompaction, or by methods such as deep dynamic compaction, vibrocompaction, or compaction grouting.

Slope Stability

Slope stability is the analysis of soil covered slopes and its potential to undergo movement. Stability is determined by the balance of shear stress and shear strength. A previously stable slope may be initially affected by preparatory factors, making the slope conditionally unstable. Triggering factors of a slope failure can be climatic events can then make a slope actively unstable, leading to mass movements. Mass movements can be caused by increases in shear stress, such as loading, lateral pressure, and transient forces. Alternatively, shear strength may be decreased by weathering, changes in pore water pressure, and organic material.

Marine Geotechnical Engineering

In subsea geotechnical engineering, seabed materials are considered a two-phase material composed of 1) rock or mineral particles and 2) water. Structures may be fixed in place in the seabed—as in piers, jettys, or fixed-bottom wind turbines—or may be floating structures anchored to remain in a sea-surface position that remain roughly fixed relative to its geotechnical anchor point.

Undersea Foundations

Figure: *Pile driving operations in the Port ofTampa, Florida, United States.*

Examples of undersea foundations include multiple-pile foundations as used in many piers and monopile foundations used for many fixed-bottom offshore wind turbines.

Floating-moored Structures

Undersea mooring of human-engineered floating structures include a large number of offshore oil and gas platforms and, since 2008, a few floating wind turbines. Two common types of engineered design for anchoring floating structures include tension-leg and catenary loose mooring systems. "Tension leg mooring systems have vertical tethers under tension providing large restoring moments in pitch and roll. Catenary mooring systems provide station keeping for an offshore structure yet provide little stiffness at low tensions." A third form of mooring system is the *ballasted catenary* configuration, created by adding multiple-tonne weights hanging from the midsection of each anchor cable in order to provide additional cable tension and therefore increase stiffness of the above-water floating structure.

Hydrogeology

Hydrogeology (*hydro-* meaning water, and *-geology* meaning the study of the Earth) is the area of geology that deals with the distribution and movement of groundwater in the soil and rocks of the Earth's crust, (commonly in aquifers). The term geohydrology is often used interchangeably. Some make the minor distinction between a hydrologist or engineer applying themselves to geology (geohydrology), and a geologist applying themselves to hydrology (hydrogeology).

Hydrogeology is an interdisciplinary subject; it can be difficult to account fully for the chemical, physical, biological and even legal interactions between soil, water, nature and society. The study of the interaction between groundwater movement and geology can be quite complex. Groundwater does not always flow in the subsurface downhill following the surface topography; groundwater follows pressure gradients (flow from high pressure to low) often following fractures and conduits in circuitous paths. Taking into account the interplay of the different facets of a multi-component system often requires knowledge in several diverse fields at both the experimental and theoretical levels. The following is a more traditional introduction to the methods and nomenclature of saturated subsurface hydrology, or simply hydrogeology.

Hydrogeology in Relation to other Fields

Hydrogeology, as stated above, is a branch of the earth sciences dealing with the flow of water through aquifers and other shallow porous media (typically less than 450 m or 1,500 ft below the land surface.) The very shallow flow of water in the subsurface (the upper 3 m or 10 ft) is pertinent to the fields of soil science, agriculture and civil engineering, as well as to hydrogeology. The general flow of fluids (water, hydrocarbons, geothermal fluids, etc.) in deeper formations is also a concern of geologists, geophysicists and petroleum geologists. Groundwater is a slow-moving, viscous fluid (with a Reynolds number less than unity); many of the empirically derived laws of groundwater flow can be alternately derived in fluid mechanics from the special case of Stokes flow (viscosity and pressure terms, but no inertial term).

The mathematical relationships used to describe the flow of water through porous media are the diffusion and Laplace equations, which have applications in many diverse fields. Steady groundwater flow (Laplace equation) has been simulated using electrical, elastic and heat conduction analogies. Transient groundwater flow is analogous to the diffusion of heat in a solid, therefore some solutions to hydrological problems have been adapted from heat transfer literature.

Traditionally, the movement of groundwater has been studied separately from surface water, climatology, and even the chemical and microbiological aspects of hydrogeology (the processes are uncoupled). As the field of hydrogeology matures, the strong interactions between groundwater, surface water, water chemistry, soil moisture and even climate are becoming more clear.

For example: Aquifer drawdown or overdrafting and the pumping of fossil water increases the total amount of water within the hydrosphere subject to transpiration and evaporation processes, thereby causing accretion in water vapour and cloud cover, the primary absorbers of infrared radiation in the earth's atmosphere. Adding water to the system has a forcing effect on the whole earth system. An accurate estimate of the climatic forcing effect due to this hydrogeological fact is yet to be quantified.

Definitions and Material Properties

One of the main tasks a hydrogeologist typically performs is the prediction of future behaviour of an aquifer system, based on analysis of past and present observations. Some hypothetical, but characteristic questions asked would be:

- Can the aquifer support another subdivision?
- Will the river dry up if the farmer doubles his irrigation?
- Did the chemicals from the dry cleaning facility travel through

the aquifer to my well and make me sick?

- Will the plume of effluent leaving my neighbour's septic system flow to my drinking water well?

Most of these questions can be addressed through simulation of the hydrologic system (using numerical models or analytic equations). Accurate simulation of the aquifer system requires knowledge of the aquifer properties and boundary conditions. Therefore a common task of the hydrogeologist is determining aquifer properties using aquifer tests. In order to further characterise aquifers and aquitards some primary and derived physical properties are introduced below. Aquifers are broadly classified as being either confined or unconfined (water table aquifers), and either saturated or unsaturated; the type of aquifer affects what properties control the flow of water in that medium (e.g., the release of water from storage for confined aquifers is related to the storativity, while it is related to the specific yield for unconfined aquifers).

Hydraulic Head

Changes in hydraulic head (h) are the driving force which causes water to move from one place to another. It is composed of pressure head ($\emptyset$) and elevation head (z). The head gradient is the change in hydraulic head per length of flowpath, and appears in Darcy's law as being proportional to the discharge.

Hydraulic head is a directly measurable property that can take on any value (because of the arbitrary datum involved in the z term); $\emptyset$ can be measured with a pressure transducer (this value can be negative, e.g., suction, but is positive in saturated aquifers), and z can be measured relative to a surveyed datum (typically the top of the well casing). Commonly, in wells tapping unconfined aquifers the water level in a well is used as a proxy for hydraulic head, assuming there is no vertical gradient of pressure. Often only *changes* in hydraulic head through time are needed, so the constant elevation head term can be left out ($\Delta h = \Delta \emptyset$). A record of hydraulic head through time at a well is a hydrograph or, the changes in hydraulic head recorded during the pumping of a well in a test are called drawdown.

Porosity

Porosity (n) is a directly measurable aquifer property; it is a fraction between 0 and 1 indicating the amount of pore space between unconsolidated soil particles or within a fractured rock. Typically, the majority of groundwater (and anything dissolved in it) moves through the porosity available to flow (sometimes called effective porosity). Permeability is an expression of the connectedness of the pores. For instance, an unfractured rock unit may have a high *porosity* (it has lots of *holes* between its constituent grains), but a low *permeability* (none of the pores are connected). An example of this phenomenon is pumice, which, when in its unfractured state, can make a poor aquifer.

Porosity does not directly affect the distribution of hydraulic head in an aquifer, but it has a very strong effect on the migration of dissolved contaminants, since it affects groundwater flow velocities through an inversely proportional relationship.

Water Content

Water content (*è*) is also a directly measurable property; it is the fraction of the total rock which is filled with liquid water. This is also a fraction between 0 and 1, but it must also be less than or equal to the total porosity.

The water content is very important in vadose zone hydrology, where the hydraulic conductivity is a strongly nonlinear function of water content; this complicates the solution of the unsaturated groundwater flow equation.

Hydraulic Conductivity

Hydraulic conductivity (K) and transmissivity (T) are indirect aquifer properties (they cannot be measured directly). T is the K integrated over the vertical thickness (b) of the aquifer ($T=Kb$ when K is constant over the entire thickness). These properties are measures of an aquifer's ability to transmit water. Intrinsic permeability (*ê*) is a secondary medium property which does not depend on the viscosity and density of the fluid (K and T are specific to water); it is used more in the petroleum industry.

Specific Storage and Specific Yield

Specific storage (S_s) and its depth-integrated equivalent, storativity ($S=S_sb$), are indirect aquifer properties (they cannot be measured directly); they indicate the amount of groundwater released from storage due to a unit depressurisation of a confined aquifer. They are fractions between 0 and 1. Specific yield (S_y) is also a ratio between 0 and 1 (S_y d” porosity) and indicates the amount of water released due to drainage from lowering the water table in an unconfined aquifer. The value for specific yield is less than the value for porosity because some water will remain in the medium even after drainage due to molecular forces. Often the porosity or effective porosity is used as an upper bound to the specific yield. Typically S_y is orders of magnitude larger than S_s.

Contaminant Transport Properties

Often we are interested in how the moving groundwater water will move dissolved contaminants around (the sub-field of contaminant hydrogeology). The contaminants can be man-made (e.g., petroleum products, nitrate or Chromium) or naturally occurring (e.g., arsenic, salinity). Besides needing to understand where the groundwater is flowing, based on the other hydrologic properties discussed above, there are additional aquifer properties which affect how dissolved contaminants move with groundwater.

Hydrodynamic Dispersion

Hydrodynamic dispersivity ($á_L$, $á_T$) is an empirical factor which quantifies how much contaminants stray away from the path of the groundwater which is carrying it. Some of the contaminants will be "behind" or "ahead" the mean groundwater, giving rise to a longitudinal dispersivity ($á_L$), and some will be "to the sides of" the pure advective groundwater flow, leading to a transverse dispersivity ($á_T$). Dispersion in groundwater is because each water "particle", passing beyond a soil particle, must choose where to go, whether left or right or up or down, so that the water "particles" (and their solute) are gradually spread in all directions around the mean path. This is the "microscopic" mechanism, on the scale of soil particles. More important, on long distances, can be the macroscopic inhomogeneities of the aquifer, which can have regions of larger or smaller permeability, so that some water can find a preferential path in one direction, some other in a different direction, so that the contaminant can be spread in a completely irregular way, like in a (three-dimensional) delta of a river.

Dispersivity is actually a factor which represents our *lack of information* about the system we are simulating. There are many small details about the aquifer which are being averaged when using a macroscopic approach (e.g., tiny beds of gravel and clay in sand aquifers), they manifest themselves as an *apparent* dispersivity. Because of this, á is often claimed to be dependent on the length scale of the problem — the dispersivity found for transport through 1 m^3 of aquifer is different than that for transport through 1 cm^3 of the same aquifer material.

Molecular Diffusion

Diffusion is a fundamental physical phenomenon, by which Einstein explained Brownian motion, which describes the random thermal movement of molecules and small particles in gases and liquids. It is an important phenomenon for small distances (it is essential for the achievement of thermodinamic equilibria), but, as the time necessary to cover a distance by diffusion is proportional to the square of the distance itself, it is ineffective for spreading a solute over macroscopic distances. The diffusion coefficient, D, is typically quite small, and its effect can often be considered negligible (unless groundwater flow velocities are extremely low, as they are in clay aquitards).

It is important not to confuse diffusion with dispersion, as the former is a physical phenomenon and the latter is an empirical factor which is cast into a similar form as diffusion, because we already know how to solve that problem.

Retardation by Adsorption

The retardation factor is another very important feature that make the motion of the contaminant to deviate from the average groundwater motion. It is analogous to the retardation factor of chromatography. Unlike diffusion and dispersion, which simply spread the contaminant, the retardation factor changes its *global average velocity*, so that it can be much slower than that of water. This is due to a chemico-physical effect: the adsorption to the soil, which holds the contaminant back and does not allow it to progress until the quantity corresponding to the chemical adsorption equilibrium has been adsorbed. This effect is particularly important for less soluble contaminants, which thus can move even hundreds or thousands times slower than water. The effect of this phenomenon is that only more soluble species can cover long distances. The retardation factor depends on the chemical nature of both the contaminant and the aquifer.

Governing Equations

Darcy's Law

Darcy's law is a Constitutive equation (empirically derived by Henri Darcy, in 1856) that states the amount of groundwater discharging through a given portion of aquifer is proportional to the cross-sectional area of flow, the hydraulic head gradient, and the hydraulic conductivity.

Groundwater Flow Equation

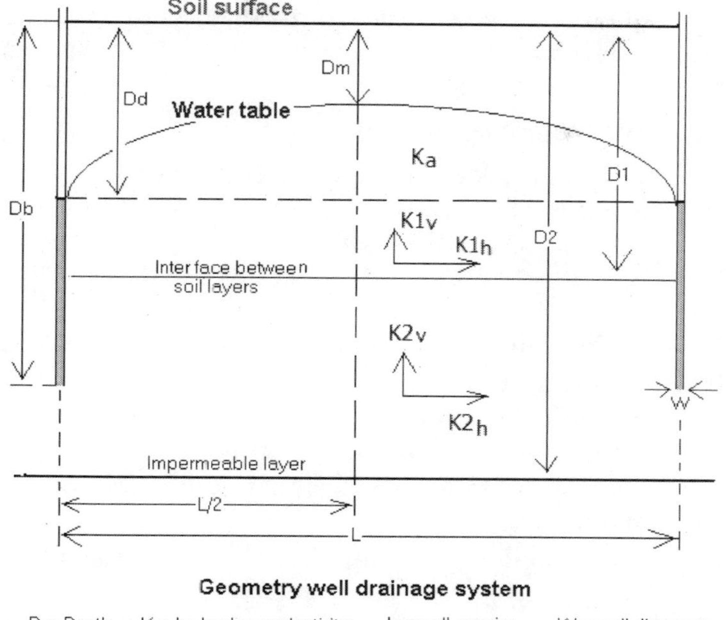

Figure: *Geometry of a partially penetrating well drainage system in an anisotropic layered aquifer*

The groundwater flow equation, in its most general form, describes the movement of groundwater in a porous medium (aquifers and aquitards). It is known in mathematics as the diffusion equation, and has many analogs in other fields. Many solutions for groundwater flow problems were borrowed or adapted from existing heat transfer solutions.

It is often derived from a physical basis using Darcy's law and a conservation of mass for a small control volume. The equation is often used to predict flow to wells, which have radial symmetry, so the flow equation is commonly solved in polar or cylindrical coordinates.

The Theis equation is one of the most commonly used and fundamental solutions to the groundwater flow equation; it can be used to predict the transient evolution of head due to the effects of pumping one or a number of pumping wells. The Thiem equation is a solution to the steady state groundwater flow equation (Laplace's Equation) for flow to a well. Unless there are large sources of water nearby (a river or lake), true steady-state is rarely achieved in reality.

Both above equations are used in aquifer tests (pump tests). The Hooghoudt equation is a groundwater flow equation applied to subsurface drainage by pipes, tile drains or ditches. An alternative subsurface drainage method is drainage by wells for which groundwater flow equations are also available.

Calculation of Groundwater Flow

To use the groundwater flow equation to estimate the distribution of hydraulic heads, or the direction and rate of groundwater flow, this partial differential equation (PDE) must be solved. The most common means of analytically solving the diffusion equation in the hydrogeology literature are:

- Laplace, Hankel and Fourier transforms (to reduce the number of dimensions of the PDE),
- similarity transform (also called the Boltzmann transform) is commonly how the Theis solution is derived,
- separation of variables, which is more useful for non-Cartesian coordinates, and
- Green's functions, which is another common method for deriving the Theis solution — from the fundamental solution to the diffusion equation in free space.

No matter which method we use to solve the groundwater flow equation, we need both initial conditions (heads at time (t) = 0) and boundary conditions (representing either the physical boundaries of the domain, or an approximation of the domain beyond that point). Often the initial conditions are supplied to a transient simulation, by a corresponding steady-state simulation (where the time derivative in the groundwater flow equation is set equal to 0).

There are two broad categories of how the (PDE) would be solved; either analytical methods, numerical methods, or something possibly in between. Typically, analytic methods solve the groundwater flow equation under a simplified set of conditions *exactly*, while numerical methods solve it under more general conditions to an *approximation*.

Analytic Methods

Analytic methods typically use the structure of mathematics to arrive at a simple, elegant solution, but the required derivation for all but the simplest domain geometries can be quite complex (involving non-standard coordinates, conformal mapping, etc.). Analytic solutions typically are also simply an equation that can give a quick answer based on a few basic parameters. The Theis equation is a very simple (yet still very useful) analytic solution to the groundwater flow equation, typically used to analyse the results of an aquifer test or slug test.

Numerical Methods

The topic of numerical methods is quite large, obviously being of use to most fields of engineering and science in general. Numerical methods have been around much longer than computers have (In the 1920s Richardson developed some of the finite difference schemes still in use today, but they were calculated by hand, using paper and pencil, by human "calculators"), but they have become very important through the availability of fast and cheap personal computers. A quick survey of the main numerical methods used in hydrogeology, and some of the most basic principles is shown below and further discussed in the article "Groundwater model".

There are two broad categories of numerical methods: gridded or discretized methods and non-gridded or mesh-free methods. In the common finite difference method and finite element method (FEM) the domain is completely gridded ("cut" into a grid or mesh of small elements). The analytic element method (AEM) and the boundary integral equation method (BIEM — sometimes also called BEM, or Boundary Element Method) are only discretized at boundaries or along flow elements (line sinks, area sources, etc.), the majority of the domain is mesh-free.

General Properties of Gridded Methods

Gridded Methods like finite difference and finite element methods solve the groundwater flow equation by breaking the problem area (domain) into many small elements (squares, rectangles, triangles, blocks, tetrahedra, etc.) and solving the flow equation for each element (all material properties are assumed constant or possibly linearly variable within an element), then linking together all the elements using conservation of mass across the boundaries between the elements (similar to the divergence theorem). This results in a system which overall approximates the groundwater flow equation, but exactly matches the boundary conditions (the head or flux is specified in the elements which intersect the boundaries).

Finite differences are a way of representing continuous differential operators using discrete intervals (Δx and Δt), and the finite difference methods are based on these (they are derived from a Taylor series). For example the first-order time derivative is often approximated using the following forward finite difference, where the subscripts indicate a discrete time location.

The forward finite difference approximation is unconditionally stable, but leads to an implicit set of equations (that must be solved using matrix methods, e.g. LU or Cholesky decomposition). The similar backwards difference is only conditionally stable, but it is explicit and can be used to "march" forward in the time direction, solving one grid node at a time (or possibly in parallel, since one node depends only on its immediate neighbours). Rather than the finite difference method, sometimes the Galerkin FEM approximation is used in space (this is different from the type of FEM often used in structural engineering) with finite differences still used in time.

Application of Finite Difference Models

Modflow is a well-known example of a general finite difference groundwater flow model. It is developed by the US Geological Survey as a modular and extensible simulation tool for modelling groundwater flow. It is free software developed, documented and distributed by the USGS. Many commercial products have grown up around it, providing graphical user interfaces to its input file based interface, and typically incorporating pre- and post-processing of user data. Many other models have been developed to work with MODFLOW input and output, making linked models which simulate several hydrologic processes possible (flow and transport models, surface water and groundwater models and chemical reaction models), because of the simple, well documented nature of MODFLOW.

Application of Finite Element Models

Finite Element programs are more flexible in design (triangular elements vs. the block elements most finite difference models use) and there are some programs available (SUTRA, a 2D or 3D density-dependent flow model by the USGS;Hydrus, a commercial unsaturated flow model; FEFLOW, a commercial modelling environment for subsurface flow, solute and heat transport processes; and COMSOL Multiphysics (FEMLAB) a commercial general modelling environment), but unless they are gaining in importance they are still not as popular in with practicing hydrogeologists as MODFLOW is. Finite element models are more popular in university and laboratory environments, where specialised models solve non-standard forms of the flow equation (unsaturated flow, density dependent flow, coupled heat and groundwater flow, etc.)

Application of Finite Volume Models

The finite volume method is a method for representing and evaluating partial differential equations as algebraic equations [LeVeque, 2002; Toro, 1999]. Similar to the finite difference method, values are calculated at discrete places on a meshed geometry. "Finite volume" refers to the small volume surrounding each node point on a mesh. In the finite volume method, volume integrals in a partial differential equation that contain a divergence term are converted to surface integrals, using the divergence theorem. These terms are then evaluated as fluxes at the surfaces of each finite volume. Because the flux entering a given volume is identical to that leaving the adjacent volume, these methods are conservative. Another advantage of the finite volume method is that it is easily formulated to allow for unstructured meshes. The method is used in many computational fluid dynamics packages.

Porflow software package is a comprehensive mathematical model for simulation of Ground Water Flow and Nuclear Waste Management developed by Analytic & Computational Research, Inc., ACRi]ACRi.

The FEHM software package is available free from Los Alamos National Laboratory and can be accessed at the FEHM Website. This versatile porous flow simulator includes capabilities to model multiphase, thermal, stress, and multicomponent reactive chemistry. Current work using this code includes simulation of methane hydrate formation, CO_2 sequestration, oil shale extraction, migration of both nuclear and chemical contaminants, environmental isotope migration in the unsaturated zone, and karst formation.

Other Methods

These include mesh-free methods like the Analytic Element Method (AEM) and the Boundary Element Method (BEM), which are closer to analytic solutions, but they do approximate the groundwater flow equation in some way. The BEM and AEM exactly solve the groundwater flow equation (perfect mass balance), while approximating the boundary conditions. These methods are more exact and can be much more elegant solutions (like analytic methods are), but have not seen as widespread use outside academic and research groups yet.

3

Reverse Engineering

Reverse engineering is the process of discovering the technological principles of a device, object, or system through analysis of its structure, function, and operation. It often involves taking something (e.g., a mechanical device, electronic component, software program, or biological, chemical, or organic matter) apart and analysing its workings in detail to be used in maintenance, or to try to make a new device or program that does the same thing without using or simply duplicating (without understanding) the original.

Reverse engineering has its origins in the analysis of hardware for commercial or military advantage. The purpose is to deduce design decisions from end products with little or no additional knowledge about the procedures involved in the original production. The same techniques are subsequently being researched for application to legacy software systems, not for industrial or defence ends, but rather to replace incorrect, incomplete, or otherwise unavailable documentation.

Motivation

Reasons for reverse engineering:

- Interoperability.
- Lost documentation: Reverse engineering often is done because the documentation of a particular device has been lost (or was never written), and the person who built it is no longer available. Integrated circuits often seem to have been designed on obsolete, proprietary systems, which means that the only way to incorporate the functionality into new technology is to reverse-engineer the existing chip and then re-design it.

- Product analysis. To examine how a product works, what components it consists of, estimate costs, and identify potential patent infringement.
- Digital update/correction. To update the digital version (e.g. CAD model) of an object to match an "as-built" condition.
- Security auditing.
- Acquiring sensitive data by disassembling and analysing the design of a system component.
- Military or commercial espionage. Learning about an enemy's or competitor's latest research by stealing or capturing a prototype and dismantling it.
- Removal of copy protection, circumvention of access restrictions.
- Creation of unlicensed/unapproved duplicates.
- Materials harvesting, sorting, or scrapping.
- Academic/learning purposes.
- Curiosity.
- Competitive technical intelligence (understand what your competitor is actually doing, versus what they say they are doing).
- Learning: learn from others' mistakes. Do not make the same mistakes that others have already made and subsequently corrected.

Reverse Engineering of Machines

As computer-aided design (CAD) has become more popular, reverse engineering has become a viable method to create a 3D virtual model of an existing physical part for use in 3D CAD, CAM, CAE or other software. The reverse-engineering process involves measuring an object and then reconstructing it as a 3D model. The physical object can be measured using 3D scanning technologies like CMMs, laser scanners, structured light digitizers, or Industrial CT Scanning (computed tomography). The measured data alone, usually represented as a point cloud, lacks topological information and is therefore often processed and modelled into a more usable format such as a triangular-faced mesh, a set of NURBS surfaces, or a CAD model.

Reverse engineering is also used by businesses to bring existing physical geometry into digital product development environments, to make a digital 3D record of their own products, or to assess competitors' products. It is used to analyse, for instance, how a product works, what it does, and what components it consists of, estimate costs, and identify potential patent infringement, etc.

Value engineering is a related activity also used by businesses. It involves de-constructing and analysing products, but the objective is to find opportunities for cost cutting.

Reverse Engineering of Software

The term *reverse engineering* as applied to software means different things to different people, prompting Chikofsky and Cross to write a paper researching the various uses and defining a taxonomy. From their paper, they state, "Reverse engineering is the process of analysing a subject system to create representations of the system at a higher level of abstraction."

It can also be seen as "going backwards through the development cycle". In this model, the output of the implementation phase (in source code form) is reverse-engineered back to the analysis phase, in an inversion of the traditional waterfall model. Reverse engineering is a process of examination only: the software system under consideration is not modified (which would make it re-engineering). Software anti-tamper technology is used to deter both reverse engineering and re-engineering of proprietary software and software-powered systems. In practice, two main types of reverse engineering emerge. In the first case, source code is already available for the software, but higher-level aspects of the program, perhaps poorly documented or documented but no longer valid, are discovered. In the second case, there is no source code available for the software, and any efforts towards discovering one possible source code for the software are regarded as reverse engineering. This second usage of the term is the one most people are familiar with. Reverse engineering of software can make use of the clean room design technique to avoid copyright infringement.

On a related note, black box testing in software engineering has a lot in common with reverse engineering. The tester usually has the API, but their goals are to find bugs and undocumented features by bashing the product from outside.

Other purposes of reverse engineering include security auditing, removal of copy protection ("cracking"), circumvention of access restrictions often present in consumer electronics, customization of embedded systems (such as engine management systems), in-house repairs or retrofits, enabling of additional features on low-cost "crippled" hardware (such as some graphics card chip-sets), or even mere satisfaction of curiosity.

Binary Software

This process is sometimes termed *Reverse Code Engineering*, or RCE. As an example, decompilation of binaries for the Java platform can be accomplished using Jad. One famous case of reverse engineering was the first non-IBM implementation of the PC BIOS which launched the historic IBM PC compatible industry that has been the overwhelmingly dominant computer hardware platform for many years. An example of a group that reverse-engineers software for enjoyment (and to distribute registration cracks) is CORE which stands for "Challenge Of Reverse Engineering".

Reverse engineering of software is protected in the U.S. by the fair use exception in copyright law. The Samba software, which allows systems that are not running Microsoft Windows systems to share files with systems that are, is a classic example of software reverse engineering, since the Samba project had to reverse-engineer unpublished information about how Windows file sharing worked, so that non-Windows computers could emulate it. The Wine project does the same thing for the Windows API, and OpenOffice.org is one party doing this for the Microsoft Office file formats. The ReactOSproject is even more ambitious in its goals, as it strives to provide binary (ABI and API) compatibility with the current Windows OSes of the NT branch, allowing software and drivers written for Windows to run on a clean-room reverse-engineered GPL free software or open-source counterpart.

Binary Software Techniques

Reverse engineering of software can be accomplished by various methods. The three main groups of software reverse engineering are

1. Analysis through observation of information exchange, most prevalent in protocol reverse engineering, which involves using bus analysers and packet sniffers, for example, for accessing a computer bus or computer network connection and revealing the traffic data thereon. Bus or network behaviour can then be analysed to produce a stand-alone implementation that mimics that behaviour. This is especially useful for reverse engineering device drivers. Sometimes, reverse engineering on embedded systems is greatly assisted by tools deliberately

introduced by the manufacturer, such as JTAG ports or other debugging means. In Microsoft Windows, low-level debuggers such as SoftICE are popular.

2. Disassembly using a disassembler, meaning the raw machine language of the program is read and understood in its own terms, only with the aid of machine-language mnemonics. This works on any computer program but can take quite some time, especially for someone not used to machine code. The Interactive Disassembler is a particularly popular tool.
3. Decompilation using a decompiler, a process that tries, with varying results, to recreate the source code in some high-level language for a program only available in machine code or bytecode.

Source Code

A number of UML tools refer to the process of importing and analysing source code to generate UML diagrams as "reverse engineering".

Reverse Engineering of Protocols

Protocols are sets of rules that describe message formats and how messages are exchanged (i.e., the protocol state-machine). Accordingly, the problem of protocol reverse-engineering can be partitioned into two subproblems; message format and state-machine reverse-engineering.

The message formats have traditionally been reverse-engineered through a tedious manual process, which involved analysis of how protocol implementations process messages, but recent research proposed a number of automatic solutions. Typically, these automatic approaches either group observed messages into clusters using various clustering analyses, or emulate the protocol implementation tracing the message processing.

There has been less work on reverse-engineering of state-machines of protocols. In general, the protocol state-machines can be learned either through a process of offline learning, which passively observes communication and attempts to build the most general state-machine accepting all observed sequences of messages, and online learning, which allows interactive generation of probing sequences of messages and listening to responses to those probing sequences. In general, offline learning of small state-machines is known to be NP-complete, while online learning can be done in polynomial time. An automatic offline approach has been demonstrated by Comparetti et al. and an online approach very recently by Cho et al.

Other components of typical protocols, like encryption and hash functions, can be reverse-engineered automatically as well. Typically, the automatic approaches trace the execution of protocol implementations and try to detect buffers in memory holding unencrypted packets.

Reverse Engineering of Integrated Circuits/Smart Cards

Reverse engineering is an invasive and destructive form of analysing a smart card. The attacker grinds away layer by layer of the smart card and takes pictures with an electron microscope. With this technique, it is possible to reveal the complete hardware and software part of the smart card. The major problem for the attacker is to bring everything into the right order to find out how everything works. Engineers try to hide keys and operations by mixing up memory positions, for example, bus scrambling. In some cases, it is even possible to attach a probe to measure voltages while the smart card is still operational. Engineers employ sensors to detect and prevent this attack. This attack is not very common because it requires a large investment in effort and special equipment that is generally only available to large chip manufacturers. Furthermore, the payoff from this attack is low since other security techniques are often employed such as shadow accounts.

Reverse Engineering for Military Applications

Reverse engineering is often used by militaries in order to copy other nations' technologies, devices, or information that have been obtained by regular troops in the fields or by intelligence operations. It was often used during the Second World War and the Cold War. Well-known examples from WWII and later include:

- Jerry can: British and American forces noticed that the Germans had gasoline cans with an excellent design. They reverse-engineered copies of those cans. The cans were popularly known as "Jerry cans".
- Tupolev Tu-4: Three American B-29 bombers on missions over Japan were forced to land in the USSR. The Soviets, who did not have a similar strategic bomber, decided to copy the B-29. Within a few years, they had developed the Tu-4, a near-perfect copy.
- V2 Rocket: Technical documents for the V2 and related technologies were captured by the Western Allies at the end of the war. Soviet and captured German engineers had to

reproduce technical documents and plans, working from captured hardware, in order to make their clone of the rocket, the R-1, which began the postwar Soviet rocket program that led to the R-7 and the beginning of the space race.

- K-13/R-3S missile (NATO reporting name AA-2 Atoll), a Soviet reverse-engineered copy of the AIM-9 Sidewinder, was made possible after a Taiwanese AIM-9B hit a Chinese MiG-17 without exploding. The missile became lodged within the airframe, and the pilot returned to base with what Russian scientists would describe as a university course in missile development.
- BGM-71 TOW Missile: In May 1975, negotiations between Iran and Hughes Missile Systems on co-production of the TOW and Maverick missiles stalled over disagreements in the pricing structure, the subsequent 1979 revolution ending all plans for such co-production. Iran was later successful in reverse-engineering the missile and are currently producing their own copy: the Toophan.
- China has reversed engineered many examples of Western and Russian hardware, from fighter aircraft to missiles and HMMWV cars.
- During the Second World War, British military intelligence at the Bletchley Park centre studied captured German "Enigma" message encryption machines. Their operation was then simulated on electro-mechanical devices called "Bombes" that tried all the possible scrambler settings of the "Enigma" machines to help break the coded messages sent by the Germans.

Legality

United States: In the United States even if an artifact or process is protected by trade secrets, reverse-engineering the artifact or process is often lawful as long as it is obtained legitimately. Patents, on the other hand, need a public disclosure of an invention, and therefore, patented items do not necessarily have to be reverse-engineered to be studied. (However, an item produced under one or more patents could also include other technology that is not patented and not disclosed.) One common motivation of reverse engineers is to determine whether a competitor's product contains patent infringements or copyright infringements.

The reverse engineering of software in the US is generally a breach of contract as most EULAs specifically prohibit it, and courts have found such contractual prohibitions to override the copyright law. Sec. 103(f) of the DMCA (17 U.S.C. § 1201 (f)) says that if you legally obtain a program that is protected, you are allowed to reverse-engineer and circumvent the protection to achieve interoperability between computer programs (i.e., the ability to exchange and make use of information). The section states:

Reverse Engineering

(1) Notwithstanding the provisions of subsection (a)(1)(A), a person who has lawfully obtained the right to use a copy of a computer program may circumvent a technological measure that effectively controls access to a particular portion of that program for the sole purpose of identifying and analysing those elements of the program that are necessary to achieve interoperability of an independently created computer program with other programs, and that have not previously been readily available to the person engaging in the circumvention, to the extent any such acts of identification and analysis do not constitute infringement under this title.

(2) Notwithstanding the provisions of subsections (a)(2) and (b), a person may develop and employ technological means to circumvent a technological measure, or to circumvent protection afforded by a technological measure, in order to enable the identification and analysis under paragraph (1), or for the purpose of enabling interoperability of an independently created computer program with other programs, if such means are necessary to achieve such interoperability, to the extent that doing so does not constitute infringement under this title.

(3) The information acquired through the acts permitted under paragraph (1), and the means permitted under paragraph (2), may be made available to others if the person referred to in paragraph (1) or (2), as the case may be, provides such information or means solely for the purpose of enabling interoperability of an independently created computer program with other programs, and to the extent that doing so does not constitute infringement under this title or violate applicable law other than this section.

(4) For purposes of this subsection, the term 0interoperability.

European Union

Article 6 of the 1991 EU Computer Programs Directive allows reverse engineering for the purposes of interoperability, but prohibits it for the purposes of creating a competing product, and also prohibits the public release of information obtained through reverse engineering of software. In 2009, the EU Computer Program Directive was superseded and the directive now states:

The unauthorised reproduction, translation, adaptation or transformation of the form of the code in which a copy of a computer program has been made available constitutes an infringement of the exclusive rights of the author. Nevertheless, circumstances may exist when such a reproduction of the code and translation of its form are indispensable to obtain the necessary information to achieve the interoperability of an independently created program with other programs. It has therefore to be considered that, in these limited circumstances only, performance of the acts of reproduction and translation by or on behalf of a person having a right to use a copy of the program is legitimate and compatible with fair practice and must therefore be deemed not to require the authorisation of the right-holder. An objective of this exception is to make it possible to connect all components of a computer system, including those of different manufacturers, so that they can work together. Such an exception to the author's exclusive rights may not be used in a way which prejudices the legitimate interests of the rightholder or which conflicts with a normal exploitation of the program.

Value Engineering

Value engineering (VE) is a systematic method to improve the "value" of goods or products and services by using an examination of function. Value, as defined, is the ratio of function to cost. Value can therefore be increased by either improving the function or reducing the cost. It is a primary tenet of value engineering that basic functions be preserved and not be reduced as a consequence of pursuing value improvements. In the United States, value engineering is specifically spelled out in Public Law 104-106, which states "Each executive agency shall establish and maintain cost-effective value engineering procedures and processes."

Value engineering is sometimes taught within the project management or industrial engineering body of knowledge as a technique in which the value of a system's outputs is optimised by crafting a mix of performance (function) and costs. In most cases this practice identifies and removes unnecessary expenditures, thereby increasing the value for the manufacturer and/or their customers.

VE follows a structured thought process that is based exclusively on "function", i.e. what something "does" not what it is. For example a screw driver that is being used to stir a can of paint has a "function" of mixing the contents of a paint can and not the original connotation of securing a screw into a screw-hole. In value engineering "functions" are always described in a two word abridgment consisting of an active verb and measurable noun (what is being done - the verb - and what it is being done to - the noun) and to do so in the most non-prescriptive way possible. In the screw driver and can of paint example, the most basic function would be "blend liquid" which is less prescriptive than "stir paint" which can be seen to limit the action (by stirring) and to limit the application (only considers paint.) This is the basis of what value engineering refers to as "function analysis".

Value engineering uses rational logic (a unique "how" - "why" questioning technique) and the analysis of function to identify relationships that increase value. It is considered a quantitative method similar to the scientific method, which focuses on hypothesis-conclusion approaches to test relationships, and operations research, which uses model building to identify predictive relationships.

Value engineering is also referred to as "value management" or "value methodology" (VM), and "value analysis" (VA). VE is above all a structured problem solving process based on function analysis—understanding something with such clarity that it can be described in two words, the active verb and measurable noun abridgement. For example, the function of a pencil is to "make marks". This then facilitates considering what else can make marks. From a spray can, lipstick, a diamond on glass to a stick in the sand, one can then clearly decide upon which alternative solution is most appropriate.

Origins

Value engineering began at General Electric Co. during World War II. Because of the war, there were shortages of skilled labour, raw materials, and component parts. Lawrence Miles, Jerry Leftow, and Harry Erlicher at G.E. looked for acceptable substitutes. They noticed that these substitutions often reduced costs, improved product, or both. What started out as an accident of necessity was turned into a systematic process. They called their technique "value analysis".

The Job Plan

Value engineering is often done by systematically following a multi-stage job plan. Larry Miles' original system was a six-step procedure which he called the "value analysis job plan." Others have varied the job plan to fit their constraints. Depending on the application, there may be four, five, six, or more stages. One modern version has the following eight steps:

1. Preparation
2. Information
3. Analysis
4. Creation
5. Evaluation
6. Development
7. Presentation
8. Follow-up.

Four basic steps in the job plan are:

- Information gathering - This asks what the requirements are for the object. Function analysis, an important technique in value engineering, is usually done in this initial stage. It tries to determine what functions or performance characteristics are important. It asks questions like; What does the object do? What must it do? What should it do? What could it do? What must it not do?
- Alternative generation (creation) - In this stage value engineers ask; What are the various alternative ways of meeting requirements? What else will perform the desired function?
- Evaluation - In this stage all the alternatives are assessed by evaluating how well they meet the required functions and how great will the cost savings be.
- Presentation - In the final stage, the best alternative will be chosen and presented to the client for final decision.

How it Works

VE follows a structured thought process to evaluate options as follows.

Gather Information

1. What is being done now?
   * Who is doing it?
   * What could it do?
   * What must it not do?

Measure

2. How will the alternatives be measured?
   * What are the alternate ways of meeting requirements?
   * What else can perform the desired function?

Analyse

3. What must be done?
   * What does it cost?

Generate

4. What else will do the job?

Evaluate

5. Which Ideas are the best?
6. Develop and expand ideas
   * What are the impacts?
   * What is the cost?
   * What is the performance?
7. Present ideas
   * Sell alternatives.

3D Scanner

A 3D scanner is a device that analyses a real-world object or environment to collect data on its shape and possibly its appearance (i.e. colour). The collected data can then be used to construct digital, three dimensional models.

Many different technologies can be used to build these 3D scanning devices; each technology comes with its own limitations, advantages and costs. Many limitations in the kind of objects that can be digitized are still present: for example optical technologies encounter many difficulties with shiny, mirroring or transparent objects.

Collected 3D data is useful for a wide variety of applications. These devices are used extensively by the entertainment industry in the production of movies and video games. Other common applications of this technology include industrial design, orthotics and prosthetics, reverse engineering and prototyping, quality control/inspection and documentation of cultural artifacts.

Functionality

The purpose of a 3D scanner is usually to create a point cloud of geometric samples on the surface of the subject. These points can then be used to extrapolate the shape of the subject (a process called reconstruction). If colour information is collected at each point, then the colours on the surface of the subject can also be determined.

3D scanners are very analogous to cameras. Like cameras, they have a cone-like field of view, and like cameras, they can only collect information about surfaces that are not obscured.

While a camera collects colour information about surfaces within its field of view, a 3D scanner collects distance information about surfaces within its field of view. The "picture" produced by a 3D scanner describes the distance to a surface at each point in the picture. This allows the three dimensional position of each point in the picture to be identified.

For most situations, a single scan will not produce a complete model of the subject. Multiple scans, even hundreds, from many different directions are usually required to obtain information about all sides of the subject. These scans have to be brought in a common reference system, a process that is usually called *alignment* or *registration*, and then merged to create a complete model. This whole process, going from the single range map to the whole model, is usually known as the 3D scanning pipeline.

Technology

There are a variety of technologies for digitally acquiring the shape of a 3D object. A well established classification divides them into two types: contact and non-contact 3D scanners. Non-contact 3D scanners can be further divided into two main categories, active scanners and passive scanners. There are a variety of technologies that fall under each of these categories.

Contact

Contact 3D scanners probe the subject through physical touch, while the object is in contact with or resting on a precision flat surface plate, ground and polished to a specific maximum of surface roughness. Where the object to be scanned is not flat or can not rest stably on a flat surface, it is supported and held firmly in place by a fixture.

The scanner mechanism may have three different forms:

- A carriage system with rigid arms held tightly in perpendicular relationship and each axis gliding along a track. Such systems work best with flat profile shapes or simple convex curved surfaces.

- An articulated arm with rigid bones and high precision angular sensors. The location of the end of the arm involves complex math calculating the wrist rotation angle and hinge angle of each joint. This is ideal for probing into crevasses and interior spaces with a small mouth opening.
- A combination of both methods may be used, such as an articulated arm suspended from a travelling carriage, for mapping large objects with interior cavities or overlapping surfaces.

A CMM (coordinate measuring machine) is an example of a contact 3D scanner. It is used mostly in manufacturing and can be very precise. The disadvantage of CMMs though, is that it requires contact with the object being scanned. Thus, the act of scanning the object might modify or damage it.

This fact is very significant when scanning delicate or valuable objects such as historical artifacts. The other disadvantage of CMMs is that they are relatively slow compared to the other scanning methods. Physically moving the arm that the probe is mounted on can be very slow and the fastest CMMs can only operate on a few hundred hertz. In contrast, an optical system like a laser scanner can operate from 10 to 500 kHz.

Other examples are the hand driven touch probes used to digitize clay models in computer animation industry.

Non-contact Active

Active scanners emit some kind of radiation or light and detect its reflection in order to probe an object or environment. Possible types of emissions used include light, ultrasound or x-ray.

Time-of-flight

The time-of-flight 3D laser scanner is an active scanner that uses laser light to probe the subject. At the heart of this type of scanner is a time-of-flight laser rangefinder. The laser rangefinder finds the distance of a surface by timing the round-trip time of a pulse of light. A laser is used to emit a pulse of light and the amount of time before the reflected light is seen by a detector is timed. Since the speed of light c is known, the round-trip time determines the travel distance of the light, which is twice the distance between the scanner and the surface. If t is the round-trip time, then distance is equal to $(C \cdot t)/2$. The accuracy of a time-of-flight 3D laser scanner depends on how precisely we can measure the *t* time: 3.3 picoseconds (approx.) is the time taken for light to travel 1 millimetre.

The laser rangefinder only detects the distance of one point in its direction of view. Thus, the scanner scans its entire field of view one point at a time by changing the range finder's direction of view to scan different points. The view direction of the laser rangefinder can be changed either by rotating the range finder itself, or by using a system of rotating mirrors. The latter method is commonly used because mirrors are much lighter and can thus be rotated much faster and with greater accuracy. Typical time-of-flight 3D laser scanners can measure the distance of 10,000~100,000 points every second.

Time-of-flight devices are also available in a 2D configuration. This is referred to as a Time-of-flight camera.

Triangulation

The triangulation 3D laser scanners are also active scanner that use laser light to probe the environment. With respect to time-of-flight 3D laser scanner the triangulation laser shines a laser on the subject and exploits a camera to look for the location of the laser dot. Depending on how far away the laser strikes a surface, the laser dot appears at different places in the camera's field of view. This technique is called triangulation because the laser dot, the camera and the laser emitter form a triangle. The length of one side of the triangle, the distance between the camera and the laser emitter is known. The angle of the laser emitter corner is also known. The angle of the camera corner can be determined by looking at the location of the laser dot in the camera's field of view. These three pieces of information fully determine the shape and size of the triangle and gives the location of the laser dot corner of the triangle. In most cases a laser stripe, instead of a single laser dot, is swept across the object to speed up the acquisition process. The National Research Council of Canada was among the first institutes to develop the triangulation based laser scanning technology in 1978.

Strengths and Weaknesses

Time-of-flight and triangulation range finders each have strengths and weaknesses that make them suitable for different situations. The advantage of time-of-flight range finders is that they are capable of operating over very long distances, on the order of kilometres. These scanners are thus suitable for scanning large structures like buildings or geographic features. The disadvantage of time-of-flight range finders is their accuracy. Due to the high speed of light, timing the round-trip time is difficult and the accuracy of the distance measurement is relatively low, on the order of millimetres. Triangulation range finders are exactly the opposite. They have a limited range of some metres, but their accuracy is relatively high. The accuracy of triangulation range finders is on the order of tens of micrometres.

Time of flight scanners accuracy can be lost when the laser hits the edge of an object because the information that is sent back to the scanner is from two different locations for one laser pulse. The coordinate relative to the scanners position for a point that has hit the edge of an object will be calculated based on an average and therefore will put the point in the wrong place. When using a high resolution scan on an object the chances of the beam hitting an edge are increased and the resulting data will show noise just behind the edges of the object. Scanners with a smaller beam width will help to solve this problem but will be limited by range as the beam width will increase over distance. Software can also help by determining that the first object to be hit by the laser beam should cancel out the second.

At a rate of 10,000 sample points per second, low resolution scans can take less than a second, but high resolution scans, requiring millions of samples, can take minutes for some time-of-flight scanners. The problem this creates is distortion from motion. Since each point is sampled at a different time, any motion in the subject or the scanner will distort the collected data. Thus, it is usually necessary to mount both the subject and the scanner on stable platforms and minimise vibration. Using these scanners to scan objects in motion is very difficult.

Recently, there has been research on compensating for distortion from small amounts of vibration.

When scanning in one position for any length of time slight movement can occur in the scanner position due to changes in temperature. If the scanner is set on a tripod and there is strong sunlight on one side of the scanner then that side of the tripod will expand and slowly distort the scan data from one side to another. Some laser scanners have level compensators built into them to counteract any movement of the scanner during the scan process.

Conoscopic Holography

In a Conoscopic system, a laser beam is projected onto the surface and then the immediate reflection along the same ray-path are put through a conoscopic crystal and projected onto a CCD. The result is a diffraction pattern, that can be frequency analysed to determine the distance to the measured surface. The main advantage with Conoscopic Holography is that only a single ray-path is needed for measuring, thus giving an opportunity to measure for instance the depth of a finely drilled hole.

Hand-held Laser Scanners

Hand-held laser scanners create a 3D image through the triangulation mechanism described above: a laser dot or line is projected onto an object from a hand-held device and a sensor (typically a charge-coupled device or position sensitive device) measures the distance to the surface. Data is collected in relation to an internal coordinate system and therefore to collect data where the scanner is in motion the position of the scanner must be determined. The position can be determined by the scanner using reference features on the surface being scanned (typically adhesive reflective tabs) or by using an external tracking method. External tracking often takes the form of a laser tracker (to provide the sensor position) with integrated camera (to determine the orientation of the scanner) or a photogrammetric solution using 3 or more cameras providing the complete Six degrees of freedom of the scanner. Both techniques tend to use infrared Light-emitting diodes attached to the scanner which are seen by the camera(s) through filters providing resilience to ambient lighting.

Data is collected by a computer and recorded as data points within Three-dimensional space, with processing this can be converted into a triangulated mesh and then a Computer-aided design model, often as Nonuniform rational B-spline surfaces. Hand-held laser scanners can combine this data with passive, visible-light sensors—which capture surface textures and colours—to build (or "reverse engineer") a full 3D model.

Structured Light

Structured-light 3D scanners project a pattern of light on the subject and look at the deformation of the pattern on the subject. The pattern may be one dimensional or two dimensional. An example of a one dimensional pattern is a line. The line is projected onto the subject using either an LCD projector or a sweeping laser. A camera, offset slightly from the pattern projector, looks at the shape of the line and uses a technique similar to triangulation to calculate the distance of every point on the line. In the case of a single-line pattern, the line is swept across the field of view to gather distance information one strip at a time.

An example of a two-dimensional pattern is a grid or a line stripe pattern. A camera is used to look at the deformation of the pattern, and an algorithm is used to calculate the distance at each point in the pattern. Consider an array of parallel vertical laser stripes sweeping horizontally across a target. In the simplest case, one could analyse an image and assume that the left-to-right sequence of stripes reflects the sequence of the lasers in the array, so that the leftmost image stripe is the first laser, the next one is the second laser, and so on. In non-trivial targets having holes, occlusions, and rapid depth changes, however, this sequencing breaks down as stripes are often hidden and may even appear to change order, resulting in laser stripe ambiguity. This problem can be solved using algorithms for multistripe laser triangulation. Structured-light scanning is still a very active area of research with many research papers published each year.

The advantage of structured-light 3D scanners is speed. Instead of scanning one point at a time, structured light scanners scan multiple points or the entire field of view at once. This reduces or eliminates the problem of distortion from motion. Some existing systems are capable of scanning moving objects in real-time.

A real-time scanner a using digital fringe projection and phase-shifting technique (a various structured light method) was developed, to capture, reconstruct, and render high-density details of dynamically deformable objects (such as facial expressions) at 40 frames per second. Recently, another scanner is developed. Different patterns can be applied to this system. The frame rate for capturing and data processing achieves 120 frames per second. It can also scan isolated surfaces, for example two moving hands.

Modulated Light

Modulated light 3D scanners shine a continually changing light at the subject. Usually the light source simply cycles its amplitude in a sinusoidal pattern. A camera detects the reflected light and the amount the pattern is shifted by determines the distance the light travelled. Modulated light also allows the scanner to ignore light from sources other than a laser, so there is no interference.

Volumetric Techniques

Medical

Computed tomography (CT) is a medical imaging method which generates a three-dimensional image of the inside of an object from a large series of two-dimensional X-ray images, similarly Magnetic resonance imaging is another a medical imaging technique that provides much greater contrast between the different soft tissues of the body than computed tomography (CT) does, making it especially useful in neurological (brain), musculoskeletal, cardiovascular, and oncological (cancer) imaging. These techniques produce a discrete 3D volumetric representation that can be directly visualised, manipulated or converted to traditional 3D surface by mean of isosurface extraction algorithms.

Industrial

Although most common in medicine, Computed tomography, Microtomography and MRI are also used in other fields for acquiring a digital representation of an object and its interior, such as nondestructive materials testing, reverse engineering, or the study biological and paleontological specimens.

Non-contact Passive

Passive scanners do not emit any kind of radiation themselves, but instead rely on detecting reflected ambient radiation. Most scanners of this type detect visible light because it is a readily available ambient radiation. Other types of radiation, such as infrared could also be used. Passive methods can be very cheap, because in most cases they do not need particular hardware but simple digital cameras.

- *Stereoscopic* systems usually employ two video cameras, slightly apart, looking at the same scene. By analysing the slight differences between the images seen by each camera, it is possible to determine the distance at each point in the images. This method is based on the same principles driving human stereoscopic vision.
- *Photometric* systems usually use a single camera, but take multiple images under varying lighting conditions. These techniques attempt to invert the image formation model in order to recover the surface orientation at each pixel.
- *Silhouette* techniques use outlines created from a sequence of photographs around a three-dimensional object against a well contrasted background. These silhouettes are extruded and intersected to form the visual hull approximation of the object. With these approaches some concavities of an object (like the interior of a bowl) cannot be detected.

User Assisted (Image-based Modelling)

There are other methods that, based on the user assisted detection and identification of some features and shapes on a set of different pictures of an object are able to build an approximation of the object itself. This kind of techniques are useful to build fast approximation of simple shaped objects like buildings. Various commercial packages are available like D-Sculptor, iModeller, Autodesk Image Modeler or PhotoModeler.

This sort of 3D scanning is based on the principles of photogrammetry. It is also somewhat similar in methodology to panoramic photography, except that the photos are taken of one object on a three-dimensional space in order to replicate it instead of taking a series of photos from one point in a three-dimensional space in order to replicate the surrounding environment.

Reconstruction

From Point Clouds: The point clouds produced by 3D scanners can be used directly for measurement and visualisation in the architecture and construction world.

Most applications, however, use instead polygonal 3D models, NURBS surface models, or editable feature-based CAD models (aka Solid models).

- Polygon mesh models: In a polygonal representation of a shape, a curved surface is modelled as many small faceted flat surfaces (think of a sphere modelled as a disco ball). Polygon models—also called Mesh models, are useful for visualisation, for some CAM (i.e., machining), but are generally "heavy" (i.e., very large data sets), and are relatively un-editable in this form. Reconstruction to polygonal model involves finding and connecting adjacent points with straight lines in order to create a continuous surface. Many applications, both free and non free, are available for this purpose (e.g. MeshLab, kubit PointCloud for AutoCAD, JRC 3D Reconstructor, imagemodel, PolyWorks, Rapidform, Geomagic, Imageware, Rhino etc.).
- Surface models: The next level of sophistication in modelling involves using a quilt of *curved* surface patches to model our shape. These might be NURBS, TSplines or other curved representations of curved topology. Using NURBS, our sphere is a true mathematical sphere. Some applications offer patch layout by hand but the best in class offer both automated patch layout and manual layout. These patches have the advantage of being lighter and more manipulable when exported to CAD. Surface models are somewhat editable, but only in a sculptural sense of pushing and pulling to deform the surface. This

representation lends itself well to modelling organic and artistic shapes. Providers of surface modellers include Rapidform, Geomagic, Rhino, Maya, T Splines etc.

- Solid CAD models: From an engineering/manufacturing perspective, the ultimate representation of a digitized shape is the editable, parametric CAD model. After all, CAD is the common "language" of industry to describe, edit and maintain the shape of the enterprise's assets. In CAD, our sphere is described by parametric features which are easily edited by changing a value (e.g., centrepoint and radius).

These CAD models describe not simply the envelope or shape of the object, but CAD models also embody the "design intent" (i.e., critical features and their relationship to other features). An example of design intent not evident in the shape alone might be a brake drum's lug bolts, which must be concentric with the hole in the centre of the drum. This knowledge would drive the sequence and method of creating the CAD model; a designer with an awareness of this relationship would not design the lug bolts referenced to the outside diametre, but instead, to the centre. A modeller creating a CAD model will want to include both Shape and design intent in the complete CAD model. Vendors offer different approaches to getting to the parametric CAD model.

Some export the NURBS surfaces and leave it to the CAD designer to complete the model in CAD (e.g., Geomagic, Imageware, Rhino). Others use the scan data to create an editable and verifiable feature based model that is imported into CAD with full feature tree intact, yielding a complete, native CAD model, capturing both shape and design intent (e.g. Geomagic, Rapidform).

Still other CAD applications are robust enough to manipulate limited points or polygon models within the CAD environment (e.g., Catia).

From a Set of 3D Slices

CT, industrial CT, MRI, or Micro-CT scanners do not produce point clouds but a set of 2D slices (each termed a "tomogram") which are then 'stacked together' to produce a 3D representation. There are several ways to do this depending on the output required:

- Volume rendering: Different parts of an object usually have different threshold values or greyscale densities. From this, a 3-dimensional model can be constructed and displayed on screen. Multiple models can be constructed from various different thresholds, allowing different colours to represent each component of the object. Volume rendering is usually only used for visualisation of the scanned object.

- Image segmentation: Where different structures have similar threshold/greyscale values, it can become impossible to separate them simply by adjusting volume rendering parameters. The solution is called segmentation, a manual or automatic procedure that can remove the unwanted structures from the image. Image segmentation software usually allows export of the segmented structures in CAD or STL format for further manipulation.
- Image-based meshing: When using 3D image data for computational analysis (e.g. CFD and FEA), simply segmenting the data and meshing from CAD can become time consuming, and virtually intractable for the complex topologies typical of image data. The solution is called image-based meshing, an automated process of generating an accurate and realistic geometrical description of the scan data.

Applications

Material Processing and Production: *Laser scanning* describes the general method to sample or scan a surface using laser technology. Several areas of application exist that mainly differ in the power of the lasers that are used, and in the results of the scanning process. Low laser power is used when the scanned surface doesn't have to be influenced, e.g. when it only has to be digitized. Confocal or 3D laser scanning are methods to get information about the scanned surface. Another low-power application are structured light projection systems that are used for solar cell flatness metrology enabling stress calculation with throughput in excess of 2000 wafers per hour. For high laser power, the influence on a working piece depends on the power of the laser: medium power values are used for laser engraving, where material is partially removed by the laser. With higher powers the material becomes fluid and laser welding can be realised, or if the power is high enough to remove the material completely, then laser cutting can be performed.

Construction Industry and Civil Engineering:

- Robotic Control: e.g., a laser scanner may function as the "eye" of a robot.
- As-built drawings of Bridges, Industrial Plants, and Monuments
- Documentation of historical sites

- Site modelling and lay outing
- Quality control
- Quantity Surveys
- Freeway Redesign
- Establishing a bench mark of pre-existing shape/state in order to detect structural changes resulting from exposure to extreme loadings such as earthquake, vessel/truck impact or fire.
- Create GIS (Geographic information system) maps and Geomatics.

Benefits of 3D Scanning

3D model scanning could benefit the design process if:

- Increase effectiveness working with complex parts and shapes.
- Help with design of products to accommodate someone else's part.
- If CAD models are outdated, a 3D scan will provide an updated version
- Replacement of missing or older parts.

Entertainment

3D scanners are used by the entertainment industry to create digital 3D models for both movies and video games. In cases where a real-world equivalent of a model exists, it is much faster to scan the real-world object than to manually create a model using 3D modelling software. Frequently, artists sculpt physical models of what they want and scan them into digital form rather than directly creating digital models on a computer.

Reverse Engineering

Reverse engineering of a mechanical component requires a precise digital model of the objects to be reproduced. Rather than a set of points a precise digital model can be represented by a polygon mesh, a set of flat or curved NURBS surfaces, or ideally for mechanical components, a CAD solid model. A 3D scanner can be used to digitize free-form or gradually changing shaped components as well as prismatic geometries whereas a coordinate measuring machine is usually used only to determine simple dimensions of a highly prismatic model. These data points are then processed to create a usable digital model, usually using specialised reverse engineering software.

Cultural Heritage

There have been many research projects undertaken via the scanning of historical sites and artifacts both for documentation and analysis purposes.

The combined use of 3D scanning and 3D printing technologies allows the replication of real objects without the use of traditional plaster casting techniques, that in many cases can be too invasive for being performed on precious or delicate cultural heritage artifacts. In the side figure the gargoyle model on the left was digitally acquired by using a 3D scanner and the produced 3D data was processed using MeshLab. The resulting digital 3D model was used by a rapid prototyping machine to create a real resin replica of original object.

Michelangelo

In 1999, two different research groups started scanning Michelangelo's statues. Stanford University with a group led by Marc Levoy used a custom laser triangulation scanner built by Cyberware to scan Michelangelo's statues in Florence, notably the David, the Prigioni and the four statues in The Medici Chapel. The scans produced a data point density of one sample per 0.25 mm, detailed enough to see Michelangelo's chisel marks. These detailed scans produced a huge amount of data (up to 32 gigabytes) and processing the data from his scans took 5 months. Approximately in the same period a research group from IBM, led by H. Rushmeier and F. Bernardini scanned the Pietà of Florence acquiring both geometric and colour details. The digital model, result of the Stanford scanning campaign, was thoroughly used in the 2004 subsequent restoration of the statue.

Monticello

In 2002, David Luebke, et al. scanned Thomas Jefferson's Monticello. A commercial time of flight laser scanner, the DeltaSphere 3000, was used. The scanner data was later combined with colour data from digital photographs to create the Virtual Monticello, and the Jefferson's Cabinet exhibits in the New Orleans Museum of Art in 2003. The Virtual Monticello exhibit simulated a window looking into Jefferson's Library. The exhibit consisted of a rear projection display on a wall and a pair of stereo glasses for the viewer. The glasses, combined with polarized projectors, provided a 3D effect. Position tracking hardware on the glasses allowed the display to adapt as the viewer moves around, creating the illusion that the display is actually a hole in the wall looking into Jefferson's Library. The Jefferson's

Cabinet exhibit was a barrier stereogram (essentially a non-active hologram that appears different from different angles) of Jefferson's Cabinet.

Cuneiform Tablets

In 2003, Subodh Kumar, et al. undertook the 3D scanning of ancient cuneiform tablets. Again, a laser triangulation scanner was used. The tablets were scanned on a regular grid pattern at a resolution of 0.025 mm (0.00098 in).

Kasubi Tombs

A 2009 CyArk 3D scanning project at Uganda's historic Kasubi Tombs, a UNESCO World Heritage Site, using a Leica HDS 4500, produced detailed architectural models of Muzibu Azaala Mpanga, the main building at the complex and tomb of the Kabakas (Kings) of Uganda. A fire on March 16, 2010, burned down much of the Muzibu Azaala Mpanga structure, and reconstruction work is likely to lean heavily upon the dataset produced by the 3D scan mission.

Figure: *Photo overlaid atop laser scan data from a project held in early 2009 at Uganda'sKasubi Tombs, which were destroyed by fire in early 2010; the data is slated for use in the building's reconstruction*

"Plastico di Roma antica"

In 2005, Gabriele Guidi, et al. scanned the "Plastico di Roma antica", a model of Rome created in the last century. Neither the triangulation method, nor the time of flight method satisfied the requirements of this project because the item to be scanned was both large and contained small details. They found though, that a modulated light scanner was able to provide both the ability to scan an object the size of the model and the accuracy that was needed. The modulated light scanner was supplemented by a triangulation scanner which was used to scan some parts of the model.

Medical CAD/CAM

3D scanners are used in order to capture the 3D shape of a patient in orthotics and dentistry. It gradually supplants tedious plaster cast. CAD/CAM software are then used to design and manufacture the orthosis, prosthesis or dental implants.

Many Chairside dental CAD/CAM systems and Dental Laboratory CAD/CAM systems use 3D Scanner technologies to capture the 3D surface of a dental preparation (either *in vivo* or *in vitro*), in order to produce a restoration digitally using CAD software and ultimately produce the final restoration using a CAM technology (such as a CNC milling machine, or 3D printer). The chairside systems are designed to facilitate the 3D scanning of a preparation *in vivo* and produce the restoration (such as a Crown, Onlay, Inlay or Veneer).

Quality Assurance and Industrial Metrology

The digitalization of real-world objects is of vital importance in various application domains. This method is especially applied in industrial quality assurance to measure the geometric dimension accuracy. Industrial processes such as assembly are complex, highly automated and typically based on CAD (Computer Aided Design) data. The problem is that the same degree of automation is also required for quality assurance. It is, for example, a very complex task to assemble a modern car, since it consists of many parts that must fit together at the very end of the production line. The optimal performance of this process is guaranteed by quality assurance systems. Especially the geometry of the metal parts must be checked in order to assure that they have the correct dimensions, fit together and finally work reliably.

Within highly automated processes, the resulting geometric measures are transferred to machines that manufacture the desired objects. Due to mechanical uncertainties and abrasions, the result may differ from its digital nominal. In order to automatically capture and evaluate these deviations, the manufactured part must be digitized as well. For this purpose, 3D scanners are applied to generate point samples from the object's surface which are finally compared against the nominal data .

The process of comparing 3D data against a CAD model is referred to as CAD-Compare, and can be a useful technique for applications such as determining wear patterns on molds and tooling, determining accuracy of final build, analysing gap and flush, or analysing highly complex sculpted surfaces. At present, laser triangulation scanners, structured light and contact scanning are the predominant technologies employed for industrial purposes, with contact scanning remaining the slowest, but overall most accurate option.

Clean Room Design

Clean room design (also known as the Chinese wall technique) is the method of copying a design by reverse engineering and then recreating it without infringing any of the copyrights and trade secrets associated with the original design. Clean room design is useful as a defence against copyright and trade secret infringement because it relies on independent invention. However, because independent invention is not a defence against patents, clean room designs typically cannot be used to circumvent patent restrictions.

The term implies that the design team works in an environment that is 'clean', or demonstrably uncontaminated by any knowledge of the proprietary techniques used by the competitor.

Typically, a clean room design is done by having someone examine the system to be reimplemented and having this person write a specification. This specification is then reviewed by a lawyer to ensure that no copyrighted material is included. The specification is then implemented by a team with no connection to the original examiners.

Examples

A famous example is that of Columbia Data Products who built the first clone of an IBM computer through a clean room implementation of its BIOS. Another is VTech's successful clones of the Apple II ROMs for the Laser 128, the only computer model, among dozens of Apple II compatibles, which survived litigation brought by Apple Computer. ReactOS is an open source operating system made from clean room reverse engineered components of Windows.

Case Law

Sony Computer Entertainment, Inc. v. Connectix Corporation was a 1999 lawsuit which established an important precedent in regard to reverse engineering. Sony sought damages for copyright infringement over Connectix's Virtual Game Station emulator, alleging that its proprietary BIOS code had been copied into Connectix's product without permission. Sony won the initial judgment, but the ruling was overturned on appeal. Sony eventually purchased the rights to Virtual Game Station to prevent its further sale and development. This established a precedent addressing the legal implications of commercial reverse engineering efforts. During production, Connectix unsuccessfully attempted a Chinese wall approach to reverse engineer the BIOS, so its engineers disassembled the object code directly. Connectix's successful appeal maintained that the direct disassembly and observation of proprietary code was necessary because there was no other way to determine its behaviour. From the ruling:

Some works are closer to the core of intended copyright protection than others. Sony's BIOS lay at a distance from the core because it contains unprotected aspects that cannot be examined without copying. The court of appeal therefore accorded it a lower degree of protection than more traditional literary works.

Code Morphing

Code morphing is one of the approaches to protect software applications from reverse engineering, analysis, modifications, and cracking used in obfuscating software. This technology protects intermediate level code such as compiled from Java and.NET languages (Oxygene, C#, Visual Basic, etc.) rather than binary object code. Code morphing breaks up the protected code into several processor commands or small command snippets and replaces them by others, while maintaining the same end result. Thus the protector obfuscates the code at the intermediate level. Code morphing is a multilevel technology containing hundreds of unique code transformation patterns. In addition this technology transforms some intermediate layer commands into virtual machine commands (like p-code). Code morphing does not protect against runtime tracing, which can reveal the execution logic of any protected code.

Unlike other code protectors, there is no concept of code decryption with this method. Protected code blocks are always in the executable state, and they are executed (interpreted) as transformed code. The original intermediate code is absent to a certain degree, but deobfuscation can still give a clear view of the original code flow.

Code morphing is also used to refer to the Just-in-time compilation technology used in Transmeta processors such as the Crusoe and Efficeon to implement the X86 instruction set architecture.

Code morphing is often used in obfuscating the copy protection or other checks that a program makes to determine whether it is a valid, authentic installation, or a pirated copy, in order to make the removal of the copy-protection code more difficult than would otherwise be the case.

4

Coordinate Measuring Machine

A coordinate measuring machine (CMM) is a device for measuring the physical geometrical characteristics of an object. This machine may be manually controlled by an operator or it may be computer controlled. Measurements are defined by a probe attached to the third moving axis of this machine. Probes may be mechanical, optical, laser, or white light, amongst others.

Description

The typical "bridge" CMM is composed of three axes, an X, Y and Z. These axes are orthogonal to each other in a typical three dimensional coordinate system. Each axis has a scale system that indicates the location of that axis. The machine will read the input from the touch probe, as directed by the operator or programmer. The machine then uses the X,Y,Z coordinates of each of these points to determine size and position with micrometre precision typically.

A coordinate measuring machine (CMM) is also a device used in manufacturing and assembly processes to test a part or assembly against the design intent. By precisely recording the X, Y, and Z coordinates of the target, points are generated which can then be analysed via regression algorithms for the construction of features. These points are collected by using a probe that is positioned manually by an operator or automatically via Direct Computer Control (DCC). DCC CMMs can be programmed to repeatedly measure identical parts, thus a CMM is a specialised form of industrial robot.

Technical Details

Parts: Coordinate-measuring machines include three main components:

- The main structure which include three axes of motion
- Probing system
- Data collection and reduction system - typically includes a machine controller, desktop computer and application software.

Uses

They are often used for:

- Dimensional measurement
- Profile measurement
- Angularity or orientation measurement
- Depth mapping
- Digitizing or imaging
- Shaft measurement.

Features

They are offered with features like:

- Crash protection
- Offline programming
- Reverse engineering
- Shop floor suitability
- SPC software and temperature compensation.
- CAD Model import capability
- Compliance with the DMIS standard
- I++ controller compatibility.

The machines are available in a wide range of sizes and designs with a variety of different probe technologies. They can be operated manually or automatically through Direct Computer Control (DCC). They are offered in various configurations such as benchtop, free-standing, handheld and portable.

Specific Parts

Machine Body: The first CMM was developed by the Ferranti Company of Scotland in the 1950s as the result of a direct need to measure precision components in their military products, although this machine only had 2 axes. The first 3-axis models began appearing in the 1960s (DEA of Italy) and computer control debuted in the early 1970s (Sheffield of the USA). Leitz Germany subsequently produced a fixed machine structure with moving table.

In modern machines, the gantry type superstructure has two legs and is often called a bridge. This moves freely along the granite table with one leg (often referred to as the inside leg) following a guide rail attached to one side of the granite table. The opposite leg (often outside leg) simply rests on the granite table following the vertical surface contour. Air bearings are the chosen method for ensuring friction free travel. In these, compressed air is forced through a series of very small holes in a flat bearing surface to provide a smooth but controlled air cushion on which the CMM can move in a frictionless manner. The movement of the bridge or gantry along the granite table forms one axis of the XY plane. The bridge of the gantry contains a carriage which traverses between the inside and outside legs and forms the other X or Y horizontal axis. The third axis of movement (Z axis) is provided by the addition of a vertical quill or spindle which moves up and down through the centre of the carriage. The touch probe forms the sensing device on the end of the quill. The movement of the X, Y and Z axes fully describes the measuring envelope. Optional rotary tables can be used to enhance the approachability of the measuring probe to complicated workpieces. The rotary table as a fourth drive axis does not enhance the measuring dimensions, which remain 3D, but it does provide a degree of flexibility. Some touch probes are themselves powered rotary devices with the probe tip able to swivel vertically through 90 degrees and through a full 360 degree rotation.

As well as the traditional three axis machines (as pictured above), CMMs are now also available in a variety of other forms. These include CMM arms that use angular measurements taken at the joints of the arm to calculate the position of the stylus tip. Such arm CMMs are often used where their portability is an advantage over traditional fixed bed CMMs. Because CMM arms imitate the flexibility of a human arm they are also often able to reach the insides of complex parts that could not be probed using a standard three axis machine.

Mechanical Probe

In the early days of coordinate measurement mechanical probes were fitted into a special holder on the end of the quill. A very common probe was made by soldering a hard ball to the end of a shaft. This was ideal for measuring a whole range of flat, cylindrical or spherical surfaces. Other probes were ground to specific shapes, for example a quadrant, to enable measurement of special features. These probes were physically held against the workpiece with the position in space being read from a 3-Axis digital readout (DRO) or, in more advanced systems, being logged into a computer by means of a footswitch or similar device. Measurements taken by this contact method were often unreliable as machines were moved by hand and each machine operator applied different amounts of pressure on the probe or adopted differing techniques for the measurement.

A further development was the addition of motors for driving each axis. Operators no longer had to physically touch the machine but could drive each axis using a handbox with joysticks in much the same way as with modern remote controlled cars. Measurement accuracy and precision improved dramatically with the invention of the electronic touch trigger probe. The pioneer of this new probe device was David McMurtry who subsequently formed what is now Renishaw plc. Although still a contact device, the probe had a spring-loaded steel ball (later ruby ball) stylus. As the probe touched the surface of the component the stylus deflected and simultaneously sent the X.Y,Z coordinate information to the computer. Measurement errors caused by individual operators became fewer and the stage was set for the introduction of CNC operations and the coming of age of CMMs.

Optical probes are lens-CCD-systems, which are moved like the mechanical ones, and are aimed at the point of interest, instead of touching the material. The captured image of the surface will be enclosed in the borders of a measuring window, until the residue is adequate to contrast between black and white zones. The dividing curve can be calculated to a point, which is the wanted measuring point in space. The horizontal information on the CCD is 2D (XY) and the vertical position is the position of the complete probing system on the stand Z-drive (or other device component). This allows entire 3D-probing.

New Probing Systems

There are newer models that have probes that drag along the surface of the part taking points at specified intervals, known as scanning probes. This method of CMM inspection is often more accurate than the conventional touch-probe method and most times faster as well.

The next generation of scanning, known as non-contact scanning includes high speed laser single point triangulation, laser line scanning, and white light scanning, is advancing very quickly. This method uses either laser beams or white light that are projected against the surface of the part. Many thousands of points can then be taken and used to not only check size and position, but to create a 3D image of the part as well. This "point-cloud data" can then be transferred to CAD software to create a working 3D model of the part. These optical scanners often used on soft or delicate parts or to facilitate reverse engineering.

Micrometrology Probes

Probing systems for microscale metrology applications are another emerging area . There are several commercially available coordinate measuring machines (CMM) that have a microprobe integrated into the system, several speciality systems at government laboratories, and any number of university built metrology platforms for microscale metrology. Although these machines are good and in many cases excellent metrology platforms with nanometric scales their primary limitation is a reliable, robust, capable micro/nano probe. Challenges for microscale probing technologies include the need for a high aspect ratio probe giving the ability to access deep, narrow features with low contact forces so as to not damage the surface and high precision (nanometre level). Additionally microscale probes are susceptible to environmental conditions such as humidity and surface interactions such as stiction (caused by adhesion, meniscus, and/orVan der Waals forces among others). Technologies to achieve microscale probing include scaled down version of classical CMM probes, optical probes, and a standing wave probe among others. However, current optical technologies cannot be scaled small enough to measure deep, narrow feature, and optical resolution is limited by the wavelength of light. X-ray imaging provides a picture of the feature but no traceable metrology information.

Physical Principles

Optical probes and/or laser probes can be used (if possible in combination), which change CMMs to measuring microscopes or multi-sensor measuring machines. Fringe projection systems, theodolite triangulation systems or laser distant and triangulation systems are not called measuring machines, but the measuring result is the same: a space point. Laser probes are used to detect the distance between the surface and the reference point on the end of the kinematic chain (i.e.: end of the Z-drive component). This can use an interferometrical function, focus variation, light deflection or a half beam shadowing principle.

Portable Coordinate Measuring Machines

Portable CMMs are different from "traditional CMMs" in that they most commonly take the form of an articulated arm. These arms have six or seven rotary axes with rotary encoders, instead of linear axes. Portable arms are lightweight (typically less than 20 pounds) and can be carried and used nearly anywhere. The inherent trade-offs of a portable CMM are manual operation (always requires a human to use it), and overall accuracy is somewhat to much less accurate than a bridge type CMM. Certain non-repetitive applications such as reverse engineering, rapid prototyping, and large-scale inspection of low-volume parts are ideally suited for portable CMMs.

Multi-Sensor Measuring Machines

Traditional CMM technology using touch probes is today often combined with other measurement technology. This includes laser, video or white light sensors to provide what is known as multi-sensor measurement.

Structural Failure

Structural failure refers to loss of the load-carrying capacity of a component or member within a structure or of the structure itself. Structural failure is initiated when the material is stressed to its strength limit, thus causing fracture or excessive deformations. In a well-designed system, a localised failure should not cause immediate or even progressive collapse of the entire structure. Ultimate failure strength is one of the limit states that must be accounted for in structural engineering and structural design.

Notable Failures

On 24 May 1847 the new railway bridge over the river Dee collapsed as a train passed over it, with the loss of 5 lives. It was designed by Robert Stephenson, using cast iron girders reinforced with wrought iron struts. The bridge collapse was the subject of one of the first formal inquiries into a structural failure. The result of the inquiry was that the design of the structure was fundamentally flawed, as the wrought iron did not reinforce the cast iron at all, and that, owing to repeated flexing, the casting had suffered a brittle failure due to fatigue.

First Tay Rail Bridge

The Dee bridge disaster was followed by a number of cast iron bridge collapses, including the collapse of the first Tay Rail Bridge on 28 December 1879. Like the Dee bridge, the Tay collapsed when a train passed over it causing 75 people to lose their lives. The bridge failed because of poorly made cast iron, and the failure of the designer

Thomas Bouch to consider wind loading on the bridge. The collapse resulted in cast iron largely being replaced by steel construction, and a complete redesign in 1890 of the Forth Railway Bridge. As a result, the Forth Bridge was the first entirely steel bridge in the world.

First Tacoma Narrows Bridge

The 1940 collapse of the original Tacoma Narrows Bridge is sometimes characterised in physics textbooks as a classical example of resonance; although, this description is misleading. The catastrophic vibrations that destroyed the bridge were not due to simple mechanical resonance, but to a more complicated oscillation between the bridge and winds passing through it, known as aeroelastic flutter. Robert H. Scanlan, father of the field of bridge aerodynamics, wrote an article about this misunderstanding. This collapse, and the research that followed, led to an increased understanding of wind/structure interactions. Several bridges were altered following the collapse to prevent a similar event occurring again. The only fatality was 'Tubby' the dog.

I-35W Bridge

The I-35W Mississippi River bridge (officially known simply as Bridge 9340) was an eight-lane steel truss arch bridge that carried Interstate 35W across the Mississippi River in Minneapolis, Minnesota, United States. The bridge was completed in 1967, and its maintenance was performed by the Minnesota Department of Transportation. The bridge was Minnesota's fifth–busiest, carrying 140,000 vehicles daily. The bridge catastrophically failed during the evening rush hour on 1 August 2007, collapsing to the river and riverbanks beneath. Thirteen people were killed and 145 were injured. Following the collapse, the Federal Highway Administration (FHWA) advised states to inspect the 700 U.S. bridges of similar construction after a possible design flaw in the bridge was discovered, related to large steel sheets called gusset plates which were used to connect girders together in the truss structure. Officials expressed concern about many other bridges in the United States sharing the same design and raised questions as to why such a flaw would not have been discovered in over 40 years of inspections.

Buildings

Ronan Point: On 16 May 1968 the 22 storey residential tower Ronan Point in the London Borough of Newham collapsed when a relatively small gas explosion on the 18th floor caused a structural wall panel to be blown away from the building.

The tower was constructed of precast concrete, and the failure of the single panel caused one entire corner of the building to collapse. The panel was able to be blown out because there was insufficient reinforcement steel passing between the panels. This also meant that the loads carried by the panel could not be redistributed to other adjacent panels, because there was no route for the forces to follow. As a result of the collapse, building regulations were overhauled to prevent disproportionate collapse and the understanding of precast concrete detailing was greatly advanced. Many similar buildings were altered or demolished as a result of the collapse.

Oklahoma City Bombing

On 19 April 1995, the nine story concrete framed Alfred P. Murrah Federal Building in Oklahoma was struck by a huge car bomb causing partial collapse, resulting in the deaths of 168 people. The bomb, though large, caused a significantly disproportionate collapse of the structure.

The bomb blew all the glass off the front of the building and completely shattered a ground floor reinforced concrete column. At second story level a wider column spacing existed, and loads from upper story columns were transferred into fewer columns below by girders at second floor level. The removal of one of the lower story columns caused neighbouring columns to fail due to the extra load, eventually leading to the complete collapse of the central portion of the building. The bombing was one of the first to highlight the extreme forces that blast loading from terrorism can exert on buildings, and led to increased consideration of terrorism in structural design of buildings.

Versailles Wedding Hall

The Versailles wedding hall, located in Talpiot, Jerusalem, is the site of the worst civil disaster in Israel's history. At 22:43 on Thursday night, May 24, 2001 during the wedding of Keren and Asaf Dror, a large portion of the third floor of the four-story building collapsed.

World Trade Centre

In the September 11 attacks, two commercial airliners were deliberately crashed into the Twin Towers of the World Trade Centre in New York City. The impact and resulting fires caused both towers to collapse within two hours. After the impacts had severed exterior columns and damaged core columns, the loads on these columns were redistributed. The hat trusses at the top of each building played a significant role in this redistribution of the loads in the structure. The impacts dislodged some of the fireproofing from the steel, increasing its exposure to the heat of the fires. Temperatures became high enough to weaken the core columns to the point of creep and plastic deformation under the weight of higher floors. Perimetre columns and floors were also weakened by the heat of the fires, causing the floors to sag and exerting an inward force on exterior walls of the building.

Aircraft

Repeated structural failures of aircraft types occurred in 1954, when 2 de Havilland Comet C1 jet airliners crashed due to decompression caused by metal fatigue, and in 1963-4, when the vertical stabiliser on 4 Boeing B-52 bombers broke off in mid-air.

Other

Warsaw Radio Mast

On 8 August 1991 at 16:00 UTC Warsaw radio mast, the tallest man-made object ever built before the erection of Burj Khalifa collapsed as consequence of an error in exchanging the guy-wires on the highest stock. The mast first bent and then snapped at roughly half its height. It destroyed at its collapse a small mobile crane of Mostostal Zabrze. As all workers left the mast before the exchange procedures, there were no fatalities, in contrast to the similar collapse of WLBT Tower in 1997.

Hyatt Regency Walkway

On 17 July 1981, two suspended walkways through the lobby of the Hyatt Regency in Kansas City, Missouri, collapsed, killing 114 and injuring 200 people at a tea dance. The collapse was due to a late change in design, altering the method in which the rods supporting the walkways were connected to them, and inadvertently doubling the forces on the connection. The failure highlighted the need for good communication between design engineers and contractors, and rigorous checks on designs and especially on contractor-proposed design changes. The failure is a standard case study on engineering courses around the world, and is used to teach the importance of ethics in engineering.

Sustainable Engineering

Sustainable engineering is the process of using energy and resources at a rate that does not compromise the natural environment, or the ability of future generations to meet their own needs.

What Engineers can do

- Water supply
- Food production
- Housing and shelter
- Sanitation and waste management
- Energy development
- Transportation
- Industrial processing
- Development of natural resources
- Cleaning up polluted waste sites
- Siting and planning projects to reduce environmental and social impacts
- Restoring natural environments such as forests, lakes, streams, and wetlands
- Improving industrial processes to eliminate waste and reduce consumption
- Recommending the appropriate and innovative use of technology.

Accomplishments from 1992 to 2002

- The World Engineering Partnership for Sustainable Development (WEPSD) was formed and they are responsible for the following areas: redesign engineering responsibilities and ethic to sustainable development, analyse and develop a long term plan, find solution by exchanging information with partners and using new technologies, and solve the critical global environment problems, such as fresh water and climate change
- Developed environmental policies, codes of ethics, and sustainable development guidelines
- Earth Charter was restarted as a civil society initiative
- The World Bank, United Nations Environmental Program, and the Global Environment Facility joined programs for sustainable development
- Launched programs for engineering students and practicing engineers on how to apply sustainable development concepts in their work
- Developed new approaches in industrial processes.

Future Goals

- Reduce the demand for predatory priced goods such as Cambodian wood through self-use of resources, to increase the demand (for the same supply), thus increase price, thus reducing demand
- Creating a comprehensive program to identify and provide the information that engineers in developing countries require to meet energy, water, food, health, and other basic human needs
- Give education for student and practicing engineers to make them realise the importance of sustainability and become environment generalists
- Engaged in decision-making processes and performing projects
- Develop better approaches with the consideration of a project's environmental costs, impacts, and conditions throughout a project's life cycle
- Improve the education on sustainability and provide help in developing countries.

Walka Water Works

Walka Water Works is a 19th century pumping station located near Maitland, New South Wales, Australia. Originally built in 1887 to supply water to Newcastle and the lower Hunter Valley, it has since been restored and preserved and is part of Maitland City Council's *Walka Recreation and Wildlife Reserve.*

History

Waterworks: During the early 1880s, with a growing population in Newcastle, the NSW Government commenced construction of the Walka Waterworks to provide a safer water supply to the Newcastle, Maitland, Morpeth and Cessnock districts than existing storage tanks, creeks and boreholes. It was largely complete by 1887 at a cost of £170,000.

The design for the complex was undertaken by notable English engineer Sir William Clark and was a major engineering project for the area at the time. Water was pumped from the Hunter River to a reservoir along a brick tunnel approximately 6 feet (1.8 m) in diametre and 9 metres (30 ft) below ground. At full operation, it had three pumping engines (150 hp each and with a flywheel weighing 36 tonnes), two horizontal compound pumps and a triple expansion surface condensing pump engine. Water would be pumped from the river at Oakhampton up to the Walka reservoir (still visible as the Walka lake today), then pumped onwards to another reservoir 6 miles (9.7 km) away at Buttai; where it was "gravitationally distributed".

At peak operation the waterworks produced 3843 megalitres (in 1915). As the population of the area continued to increase, alterations and additions were made to the system until 1913, when other sources of supply were developed - namely Chichester Dam near Dungog. From 1923 to 1940 the waterworks were used as a backup water supply only. In 1925 the complex was put on standby and with the onset of the great depression and the completion of Tarro pumping station at Tarro, the plant was closed in 1929. After the Second World War in 1949 all the plant and machinery was sold for scrap, fetching £2,500.

Power Station

Two years later, in 1951, the site was reopened as a temporary coal power station by the Electricity Commission to overcome post-war electricity shortages. What was described as a "package power plant" was bought from General Electric and imported from America. The plant was shipped out in pieces, then taken to the site by rail (to Maitland station) and a specially made truck. GE engineers also came out to assemble the plant. It began producing electricity in 1953. Three boilers ran on coal, and another ran on oil, though later this became a coal/oil combination. Two rail lines were built to the power station from the North Coast railway line.

Using Walka as a power station was controversial. Initially the power station suffered from sinking foundations - leading to the temporary sacking of 120 workers, until it was agreed that the sinking was a result of recent rain. During construction, the workforce went on strike after a boilermaker union member was sacked for "misconduct" towards his foreman. This was eventually resolved, but in December 1953 the power station suffered another setback when one of its oil storage tanks collapsed in gale force winds. The station was also unpopular with local vegetable growers, who complained about the amount of soot the station was producing.

This station was eventually decommissioned in 1978 and the site was closed.

Recreation and Wildlife Reserve

Not until 1984 were there definite attempts to reopen the site. At this time a Trust was formed, aiming to open the site and restore the Waterworks complex. The complex had been classified by the National Trust in 1976. Today the area is open as a free public reserve, with barbecues, picnics areas, a playground, walking trails and a 7 1/4 inch gauge toy train. The reservoir and surrounding plant life make it a unique environment for birds and animals in the area.

Key Characteristics

- Main pumphouse with distinctive chimney, ornate brickwork and Victorian Italianate architecture
- Large sandstone water reservoir
- Settling tanks and filter beds to purify water
- Toy train.

Science Barge

The Science Barge is an itinerant floating science museum now docked in Yonkers, New York, USA. It is also a working urban farm operated by the sustainable development organisation Groundwork Hudson Valley, and designed by the New York Sun Works Centre for Sustainable Engineering. The Science Barge grows crops using a hydroponic greenhouse powered by solar panels, wind turbines, and biofuels. The crops in the greenhouse are irrigated by captured rainwater and desalinated river water. Food is grown without carbon emissions, no agricultural waste is discharged into the watershed and no pesticides are used. The barge was designed by the visionary environmental engineer Dr. Ted Caplow.

The goal of the Science Barge is to promote food production within cities around the world in an ecologically sustainable way. The Science Barge uses this goal to teach sustainability to students of all ages, and has an active education program for school and camp trips. The Barge is open to the public.

In November 2008, the Science Barge docked in Yonkers, New York, USA on the Hudson River at the mouth of the Saw Mill River. The Yonkers downtown site was selected for a variety of reasons, including: (1) to add yet another exciting destination to the Yonkers downtown to create a vibrant recreation and commerce district; (2) to provide easy access to mass transit including MetroNorth, Amtrak, and Bee-Line Bus System at the historic Yonkers train station, and New York Water Taxi at the Yonkers Pier; (3) to offer easy bicycle and pedestrian access from the Waterfront Promenade; and, (4) to highlight the restoration work of the Saw Mill River - a major Hudson

River tributary. At one time it was located on Pier 92, next to the New York Passenger Ship Terminal, the Circle Line Sightseeing Cruises pier, and the Intrepid Sea-Air-Space Museum. In 2007 it was at Pier 84, until the end of October, where it reappeared in late April 2008.

As awareness of sustainable food production methods and specific interest in Building-integrated agriculture (which the Barge was designed to promote) have grown, coverage of the Barge has increased. In July 2009, GOOD created a short video of the purpose and methods of the Barge. In the same month, former CBS News Anchor Dan Rather hosted an episode of "Dan Rather Reports" on the Barge. Prior coverage included a 2008 podcast with Jen Nelkin and Zak Adams of the Science Barge speaking to the New York Academy of Sciences about the barge and how it functions. Other original coverage from 2007-2009 can be found on the "New York Sun Works press page". A hydroponic greenhouse, inspired by the Science Barge, was recently opened on the roof of the Manhattan School for Children.

In March 2009, the Science Barge was named "Best Class Trip" by *New York* magazine in its annual "Best Of..." issue. In June 2010 it was docked in Yonkers.

London's Water Supply Infrastructure

London's water supply infrastructure has developed over the centuries in line with the expansion of London and now represents a sizeable infrastructure investment. For much of London's history, private companies supplied fresh water to various parts of London from the River Thames and the River Lea. A crisis point was reached in the mid 19th century with outbreaks of cholera and general problems arising from extraction of water from the polluted Tideway, and major new facilities were built up river at Hampton and Molesey. After merger and nationalisation into the Metropolitan Water Board, and later reprivatisation, their modern descendent Thames Water still runs London's water supply infrastructure.

Early London Water Supply

Until the late 16th century, London citizens relied on the River Thames, its tributaries, or one of around a dozen natural springs for their water supplies. In 1247 work began on building the Great Conduit from the spring at Tyburn. This was a lead pipe which led via Charing Cross, Strand, Fleet Street and Ludgate to a large cistern or tank in Cheapside. The city authorities appointed keepers of the conduits who controlled access so that users such as brewers, cooks and fishmongers would pay for the water they used. Wealthy Londoners living near the a conduit pipe could obtain permission for a connection to their homes, but this did not prevent unauthorised tapping of conduits. Otherwise - particularly for households which could not take a gravity-feed - water from the conduits was provided to individual households by water carriers or "cobs". Records of frequent drownings indicate that many poorer citizens collected water from the Thames or nearby streams running into the Thames. The Grand Conduit system was extended over the centuries and in the 15th century was supplemented by a conduit from springs at Paddington, and another at Highgate which supplied Cripplegate.

Sixteenth Century

In 1582, Dutchman Peter Morice (died 1588) developed one of the first pumped water supply systems for the City of London, powered by undershot waterwheels housed in the northernmost arches of London Bridge spanning the River Thames. The machinery was largely destroyed in the Great Fire of London in 1666 but replacements engineered by his grandson remained under the bridge until the early 19th century, before the New London Bridge was erected in the 1830s.

Seventeenth Century

Hugh Myddleton was the driving force behind the construction of the New River, an ambitious engineering project to bring fresh water from Hertfordshire to 17th century London. After the initial project encountered financial difficulties, Myddleton helped fund the project through to completion. The New River was constructed between 1609 and 1613 (being officially opened on 29 September that year), and was originally some 38 miles (60 km) long. It was not initially a financial success, and cost Myddleton substantial sums, although in 1612 he was successful in securing monetary assistance from King James I. The New River Company became one of the largest private water companies, supplying the City of London and other central areas.

The construction of London's current water distribution infrastructure dates back to the Great Fire of London in 1666, which destroyed most of the city's previous water infrastructure, most of which was made of wood and lead. One waterworks not affected by the fire was at Shadwell which dated from 1660. The city's water supply and distribution infrastructure has been continuously updated and upgraded since then.

Eighteenth Century

The Chelsea Waterworks Company was established in 1723 "for the better supplying the City and Liberties of Westminster and parts adjacent with water". The company received a Royal Charteron 8 March 1723. The company created extensive ponds in the area bordering Chelsea and Pimlico using water from the tidal Thames.

Waterworks were established in East London, at West Ham in 1743 and at Lea Bridge before 1767.

The Borough Waterworks Company was formed in 1770, originally supplying water to a brewery and the surrounding area, which spanned the distance between London and Southwark Bridges. The adjacent area was supplied by the London Bridge Waterworks Company.

The Lambeth Waterworks Company was founded in 1785 to supply water to south and west London. It was established on the south bank of the River Thames close to the present site of Hungerford Bridge where the Royal Festival Hall now stands. The first water intake of the company was on the south side of the river and supplied directly from the river. After complaints that the water was foul, the intake was moved to the middle of the river.

Nineteenth Century

New Companies

As London spread in the 19th century, new facilities were needed to service the increasing population in newly developed areas. Several new water supply companies were established leading to an arrangement of up to nine private water companies each with a geographic monopoly.

The Lambeth Waterworks company expanded in 1802 to supply Kennington and about this time replaced its wooden pipes with iron ones. The South London Waterworks Company was established by private act of parliament in 1805. The company extracted water from the Thames beside Vauxhall Bridge.

The West Middlesex Waterworks Company was founded in 1806 to supply water to the Marylebone and Paddington areas of London. In 1808 the company installed cast iron pipes to supply water from its intakes at Hammersmith.

The East London Waterworks Company was founded by Act of Parliament in 1806, and also acquired existing waterworks at Shadwell, Lea Bridge and West Ham.

The Grand Junction Waterworks Company was created in 1811 to take advantage of a clause in the Grand Junction Canal Company's Act which allowed them to supply water brought by the canal from the River Colne and River Brent, and from a reservoir in the north-west Middlesex supplied by land drainage. It was thought that these waters would be better than those of the Thames, but in fact they were found to be of poor quality and insufficient to meet demand. After trying to resolve these problems the company resorted to taking their supply from the River Thames at a point near Chelsea Hospital

Expansion

Although the legislation that established the London water companies intended that they would compete for customers, in 1815 the East London company drew up a legal agreement with the New River Company defining a boundary between their areas of supply.

The London Bridge Waterworks Company was dissolved in 1822, and its water supply licence was purchased by the New River Company. Later that same year, the Borough Waterworks Company purchased the London Bridge licence from the New River Company, and it was renamed the Southwark Water Company. The company extracted water from the River Thames using steam engines to pump it to a cistern at the top of a 60-foot-high (18 m) tower.

The West Middlesex Waterworks Company established a 3.5 million gallon reservoir at Campden Hill near Notting Hill. In 1825 the company built a new reservoir at Barrow Hill next to Primrose Hill in North London.

In 1825, the ponds of the Chelsea Waterwork Company were used as the basis of the Grosvenor Canal which was opened to traffic that year. By this time there were complaints about the quality of the water that the company was drawing from the River Thames, and in 1829, under engineer James Simpson the Chelsea Waterworks Company became the first in the country to install a slow sand filtration system to purify the water.

In 1829, the East London Waterworks Company moved their source of water further up river to Lea Bridge as a result of pollution caused by population growth. Clean water was now abstracted from the natural channel which had been by-passed by the Hackney Cut, to a new reservoir at Old Ford. In 1830 the company gained a lease on the existing reservoir at Clapton.

In 1832 the Lambeth Waterworks Company built a reservoir at Streatham Hill, and in 1834 obtained an Act of Parliament to extend its area of supply. In the same year, the Company purchased 16 acres (65,000 m^2) of land in Brixton and built a reservoir and works on Brixton Hill adjacent to Brixton Prison.

In 1833 the South London Waterworks Company was supplying 12,046 houses with approximately 12,000 gallons of water. In 1834, the company was renamed the Vauxhall Water Company.

The Grand Junction Waterworks Company built a pumping station near Kew Bridge at Brentford in 1838 to house its new steam pump and two similar pumps purchased from Boulton, Watt and Company in 1820. The water was taken from the middle of the river and pumped into filtering reservoirs and to a 200 ft (61 m) high water tower to provide gravity feed to the area. A six to seven mile (11 km) main took the water to a reservoir on Campden Hill near Notting Hill capable of containing 6 million gallons.

In 1841 the East London Waterworks Company was supplying 36,916 houses.

On 10 January 1845 the Southwark Waterworks Company, and the Vauxhall Waterworks Company submitted a memorandum to the Health of Towns Commissioners proposing amalgamation. The bill promoted by the two companies successfully passed through parliament, and the Southwark and Vauxhall Waterworks Company was formed later that year. The area supplied by the SVWC was centred on the Borough of Southwark, reaching east to Rotherhithe, south to Camberwell and in the west including Battersea and parts of Clapham and Lambeth. The amalgamated company established waterworks at Battersea Fields with two depositing reservoirs with a capacity of 32 million gallons; and two filtering reservoirs holding 11 million gallons. In 1850 the company's water was described by the microbiologist Arthur Hassall as "the most disgusting which I have ever examined".

In 1845 the limits of supply of the East London Waterworks Company were described as *"all those portions of the Metropolis, and its suburbs, which lie to the east of the city, Shoreditch, the Kingsland Road, and Dalston; extending their mains even across the river Lea into Essex, as far as West Ham."* The water supplied by the company was taken from the Lea, with waterworks on 30 acres (0.12 km^2) of land at Old Ford.

Metropolis Water Act

The companies often provided inadequate quantities of water which was often contaminated, as was famously discovered by John

Snow during the 1854 cholera epidemic. Population growth in London had been very rapid (more than doubling between 1800 and 1850) without an increase in infrastructure investment. The Metropolis Water Act 1852 was enacted in order to "make provision for securing the supply to the Metropolis of pure and wholesome water." Under the Act, it became unlawful for any water company to extract water for domestic use from the tidal reaches of the Thamesafter 31 August 1855, and from 31 December 1855 all such water was required to be "effectually filtered". The Metropolitan Commission of Sewers was formed, water filtration was made compulsory, and new water intakes on the Thames were established above Teddington Lock.

The Chelsea Waterworks Company and the Lambeth Waterworks Company, who shared the services of James Simpson, established facilities at Seething Wells between Thames Ditton andSurbiton. The Chelsea's former site was taken over by the railways to make space for Victoria Station. The Grand Junction, West Middlesex and Southwark and Vauxhall Waterworks Companies set up facilities above Molesey Lock at Hampton designed by Joseph Quick. The Stain Hill Reservoirs and Sunnyside Reservoir were constructed in Hampton by the SVWC in 1855, with a 36-inch (910 mm) diametre main to Battersea. A third reservoir was opened later in the year between Nunhead Cemetrey and Peckham Rye.

In the mid 19th century the East London Waterworks Company purchased the Coppermill at Walthamstow and modified it to drive a water pump to assist in the building of reservoirs on nearby marshland in the Lea Valley . The company built a series of reservoirs which were High Maynard Reservoir, Low Maynard Reservoir, five linked numbered reservoirs making the Walthamstow Reservoirs, the East Warwick Reservoir and the West Warwick Reservoir.

In 1872 the Lambeth Waterworks Company moved upstream on the Thames to Molesey, followed by the Chelsea Waterworks Company. They built the Molesey Reservoirs there in 1872.

The East London Waterworks Company replaced their reservoir at Clapton by a new reservoir at Stamford Hill in 1891.

In 1897 the New River Company started developing the treatment works at Kempton Park to supply additional water to their facilities at Cricklewood.

In 1898 the SVWC started work on the Bessborough Reservoir and the Knight Reservoir which were across the river from Hampton at Molesey. By 1903 the SVWC supplied a population of 860,173 in 128,871 houses of which 122,728 (95.3%) had a constant supply. The Lambeth Waterworks company started work on Island Barn Reservoir at Molesey in 1900.

Twentieth Century

The private water companies were nationalised at the beginning of the 20th century. The Metropolis Water Act 1902 (2 Edw.7, c.41) created the Metropolitan Water Board. It was founded in 1903 and as originally constituted in the Act had 67 members; 65 of these were nominated by local authorities, who appointed a paid chairman and vice-chairman. The board compulsorily acquired the following water companies:

- The New River Company
- The East London Waterworks Company
- The Southwark and Vauxhall Waterworks Company
- The West Middlesex Waterworks Company
- The Lambeth Waterworks Company
- The Chelsea Waterworks Company
- The Grand Junction Waterworks Company
- The Staines Reservoirs Joint Committee.

Also acquired at no cost were the water undertakings of Tottenham and Enfield Urban District Councils.

The MWB opened the East London Waterworks reservoirs Banbury Reservoir and Lockwood Reservoir, and the Bessborough Reservoir, Knight Reservoir and Island Barn Reservoirs at Molesey. It also opened the Kempton Park Reservoirs in around 1907.

In 1910, extraction facilities were opened at Hythe End and the Staines Reservoir Aqueduct was built to supply water to Hampton. The Metropolitan Water Board Railway was opened in 1916 to carry coal from the river at Hampton to Kempton Park. An engine house with powerful steam engines was opened at Kempton Park in 1929, which has now become Kempton Park Steam Engines museum.

The MWB opened a succession of reservoirs - King George V Reservoir, (Lea Valley) in 1912, Queen Mary Reservoir (Ashford) in 1925, King George VI Reservoir (Stanwellmoor) in 1947 William Girling Reservoir (Lea Valley) in 1951, Queen Elizabeth II Reservoir (Molesey) 1962, Wraysbury Reservoir 1967, and Queen Mother Reservoir (Staines)

1976. The Metropolitan Water Board and other local Water Boards were later combined into the Thames Water Authority, which was later privatised as Thames Water, a state-regulated private company which currently provides London's water supply.

Present Day

Most of London's water still comes from the River Thames and River Lea, with the remainder being abstracted from underground sources.

Much of the water piping in London is still cast iron piping which dates back to the nineteenth century and is slowly deteriorating. This has led to widespread criticism of Thames Water for the amount of water lost to leaks in its distribution network. As of 2007, Thames Water is still in the process of a rolling programme of upgrading the water supply network to use modern plastic piping.

The single largest infrastructure project in recent years has been the creation of the Thames Water Ring Main, a "backbone network" for London's water supply. This connects all the waterworks, and pumping stations.

Water Treatment Works

The Water treatment works on the Ring Main are as follows.

- Ashford Common
- Kempton Park
- Walton
- Hampton
- Coppermills Water Treatment Works.

Pumping Stations

The pumping stations on the ring main are as follows.

- Surbiton — PS
- Merton — PS
- Streatham — PS
- Brixton — PS
- Battersea — PS
- Park Lane — PS
- Kew — PS
- Holland Park Avenue — PS

- Barrow Hill — PS
- New River Head — PS
- Stoke Newington — PS.

Access

- Hogsmill — Access
- Raynes Park — Access
- Mogden — Access.

Storage

- Honour Oak — Underground Storage
- Barnes — balancing storage.

Chaplin's Patent Distilling Apparatus with Steam Pump

Alexander Chaplin & Co. (also known as A.C. and Co., Alex. Chaplin & Co.) was an important and highly regarded engineering and manufacturing syndicate based in the United Kingdom during the mid-19th century to early 20th century with its manufacturing and plant facilities located at Cranstonhill Engine Works, Glasgow. Of their numerous patented and manufactured products, one of the notable devices developed was the Chaplin's Patent Distilling Apparatus with Steam pump for circulating water attached. This was an early design of an evaporator, a device for producing fresh water on board ship by distillation of seawater. An example of this apparatus has been recovered from the wreck of *SS Xantho* (1872), a steamship used in Australia to transport passengers and trade goods before ultimately sinking in Port Gregory, Western Australia in 1872. It is purported that the Alexander Chaplin distiller from the *SS Xantho* wreck is the only known surviving example of a Chaplin distilling apparatus on board a steamship of its time.

History

Founded in 1857, the administrative office of Alexander Chaplin & Co. was located at 63 Queen Victoria Street, London, England, before being taken over by Herbert Morris, LTD in 1932. Well known manufacturers, Alex. Chaplin & Co. prided themselves as "always in stock or in progress". Chaplins' patent steam engines and boilers powered steam cranes, hoists, locomotives, pumping and winding engines, ship's deck engines and sea water distilling apparatus throughout the world during the 19th and 20th century. Chaplin's patent sea water distilling apparatus with steam pump attached for circulating water was a very compact and convenient apparatus, used not only for seafaring purposes but on land as well.

The Chaplin Apparatus was adopted by many important British and Continental shipping companies including the Peninsular and Oriental, the Inman, the North German Lloyd, and the Hamburg American Companies. The British Navy used such apparatus'. The Board of Bureau Chiefs of the United States Navy considered supplying the cruisers *USS Baltimore (C-3)* and *USS Philadelphia (C-4)* with distilling apparatus to provide fresh water for the boilers from distillation of sea water, but no reference was made to the Chaplin company in their reported discussions. In addition to merchant and passenger ships US gunboats used such distillers as well.

Chaplin distillers were also used in land settings. During the Sudan Campaigns (1881–1885), (1896–1899), British forces used the distillers to supply their soldiers with fresh drinking water in Suakinand Sudan. This supply of fresh drinking water was of utmost importance in the 1882 Anglo-Egyptian War. Similar distillers, such as Dr. Normandy's, were on a much larger scale, distilling some 12,000 gallons of water per hour, whereas the Chaplin distilling apparatus used onboard vessels produced roughly 23 gallons of fresh water per hour. Essentially, the distilled water produced for Egypt was made in a special apparatus with various forms of condenser employed. The principle for distillation is the same as the apparatus on steamships. On ships, steam is generated in one of the ships boilers then condensed, filtered and aerated in the apparatus. On land, the engines would have to be kept running in order to pump the distilled water out of the condenser.

Figure: *Example of Chaplin Distiller used for Land Purposes*

By the order of H.M. Council on 5 December 1865, Government emigrant, troop and other passenger vessels fitted with this distilling apparatus were permitted to sail with only half the required amount of water under Section 26 of the *Passengers Act 1855*. This act states "Any passenger ship propelled by sails only, or by steam engines of less power than is sufficient, without the aid of sails, to propel the ship at the rate of five statue miles per hour, may be cleared out and proceed on her voyage, having on board, in tanks or casks, only half the quantity of pure water required by the said Act to be carried for the use of the passengers, provided that...there be on board such ship an efficient apparatus for distilling fresh water from salt water". This, combined with the convenience and compact size of the apparatus made the Alexander Chaplin model very popular.

In direct competition with Alexander Chaplin, Dr. Alphonse Normandy's (1809–1864) apparatus is another successful distiller. It consists of three essential parts, the evaporator, the condenser and the refrigerator. The apparatus passes a mixture of steam and gasses from the evaporator to the tubes of the condenser. Normandy's distiller was very complex in structure, consisting of many numerous working parts. With an elaborate layout and expensive cost, the distiller wasn't economically plausible from a moderate vessel's standpoint. In France, the apparatus of Rocher and Nantes and that of Galle and Mazeline of Havre, have been "highly appreciated by French authorities and French seamen".

How it Works

For successful preparation of potable water from seawater, the following conditions of a distilling apparatus are essential. First, the distilled product must be aerated so it may be immediately available for drinking and storage purposes. Second, the amount of coal used to obtain the maximum volume of drinkable water must be at a minimum expenditure level and third, the apparatus' working parts must be simple enough to prevent from breaking down and to enable unskilled attendants to safely operate. "The Alexander Chaplin distilling apparatus is among the forms of apparatus which have most fully satisfied such conditions".

The Apparatus

The Alexander Chaplin & Co. distilling apparatus is designed for use on board steam vessels and may be connected from the main boiler or a donkey boiler. The apparatus consists of a coiled pipe approximately 60 feet in length placed inside a cylindrical casing. The casing is cast iron and the coils are made of copper. The coils are supplied with steam by one of two methods. Either from the exhaust pipes of the engine, or from a small fitting leading directly from the boiler. A double-acting circulating pump produces a constant stream of water flowing through the condenser, using a pinion that can be disengaged from the crank shaft manually.

At the upper end of the coils lie a covered brass cup for ventilating. This aerator is the essential feature in the invention and consists of small adjustable holes around the circumference of the brass cup that allows a pipe to enter the coils. When the apparatus is in action, a jet draws air in from the circumferential openings in the cup and fills the coils with a mixture of air and steam. After passing directly under the condenser, the aerated water is delivered clear, bright and odourless at a temperature of about 13 degrees Celsius. The resulting fresh water is of excellent quality and ready for immediate use. The machine can supply approximately 23 gallons of water per hour. "The samples of water subjected to analysis were collected at different times directly from the Patent Apparatus on board ship... and the results of my experiments with the water distilled in Alexander Chaplin & Co.'s are conclusive in showing that is of excellent quality in every particular." The circulating pump of the condensing arrangement can be used as a fire engine in the event of an emergency. The waste pipe from the condenser allows for a cock to be attached, by which way the water can be directed to a fitted union which a hose can be attached. Theoretically, the water would be able to be pumped back into the boiler as well. It is preferred to place the apparatus in the Stokehole or elsewhere below deck, but may be placed on the main deck or any other convenient part of the vessel. In addition to the safety advantages offered by the Chaplin distiller by only having to carry half the amount of water on board it also provided an economic benefit in that the space and weight previously set aside for larger water tanks could be used more productively to carry additional cargo.

According to the manufacturer's specifications and a report generated by Dr. Frederick Penny Ph.D., F.R.S.E Professor of Chemistry, Andersonian University, Glasgow, this distiller filters out noxious chemical such as lead, copper, tin and iron through the use of a series of iron plates with perforated holes and matting. These plates are located in the large box like structure at the bottom of the apparatus. Precisely when the Chaplin Distiller was installed on the *SS Xantho* is open to conjecture. It is known that from 1864-1870 the *SS Xantho* operated out of Wick, West Sussex and was permitted to take excursions to sea. It is possible that this apparatus was installed during this period of time. It is just as feasible however that this device was installed at a later date as part of the major refurbishing programme implemented by the metal merchant Robert Stewart, who is known to have installed a new marine fire tube boiler and a second hand Horizontal Trunk Engine in 1871, designed by John Penn and Son.

Excavation and Confirmation of Identity

This Chaplin distiller was excavated from the steamship SS *Xantho* (1872) wreck site in 1994. It was found outside the hull on the port side of the vessel. It is possible that this apparatus was installed on the ship's deck or in its cabin and as a result of the wreck sites natural transformation has rolled off the deck onto the seabed.

Initially this artefact was considered to be a condenser or a heat exchanger perhaps circle condenser but the discovery of a brass manufacturer's nameplate found whilst de-concreting the object and subsequent archival documentation has confirmed its identity to be a Chaplin distilling apparatus.

Figure: *Distiller cannister of SS Xantho*

On recovery the artefact was found to be in very fragile condition with the cast iron shell badly corroded and incomplete. Additionally there is evidence to suggest that the artefact is missing many of its mountings such as the aerator, connecting pipes and donkey engine. It is inconclusive whether these materials were removed by agents operating under the direction of the owner Charles Edward Broadhurst (1826 – 26 April 1905) in the initial salvaging of the wreck in 1872 or by treasure hunters in more recent times.

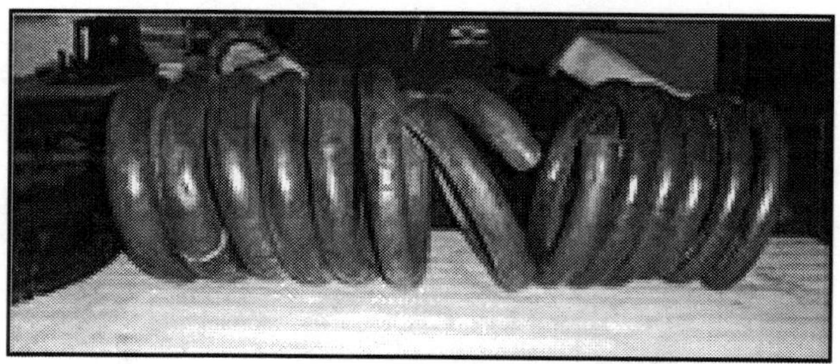

Figure: *Distiller coil from SS Xantho*

The Chaplin Distiller which was damaged during the excavation and treatment processes is currently being reconstructed at the Western Australian Museum, by the Department of Maritime Archaeology. It is envisaged that this rebuilding process will be completed by 2012 and that it will be available for public viewing in the 'Steamships to Suffragettes Gallery' shortly thereafter.

Because of its fragmented state and damage to the coil it is not possible at this stage to provide exact dimensions for this apparatus. A rough estimate would suggest something in the order of 5 foot 3 inches but a definitive answer will be able to be determined once it is reassembled and the coil has been correctly aligned and integrated. The Chaplin Distilling apparatus excavated from the *SS Xantho* is believed to be the only known example in existence and therefore can be regarded as a unique piece of maritime history that helps to explain how steamships and sailing vessels fitted with these devices were able to undertake prolonged voyages at sea with a minimal supply of fresh water.

Andreyevsky Bridge

Andreyevsky Bridge (Russian: *St.Andrew's Bridge*) name refers to a historical bridge demolished in 1998 and three existing bridges across Moskva River, located between Luzhniki and Gorky Park in Moscow.

Andreyevsky Rail Bridge (1907, Demolished 1998)

The original Sergievsky Bridge, named after the late Grand Duke Sergei Alexandrovich, and its twin, Nicholas II Bridge were built in 1903-1907 , replacing temporary wooden truss bridges of the Moscow Inner Ring Railroad.

New bridges were designed as through arch bridges by Lavr Proskuryakov (structural engineering) and Alexander Pomerantsev (architectural design). The 135 metre wide, 15 metre tall steel arch (1400 metric tons) was made at Votkinsk works. Proskuryakov's work, considered a marvel of engineering, was proven by the 1908 flood: water level exceeded the maximum design specification by a metre and a half; the bridges stood unharmed. After the February Revolution of 1917, Romanov names were erased from the map. Sergievsky Bridge was renamed *Andreyevsky* after nearby St.Andrew's monastery, Nicholas II Bridge became Krasnoluzhsky Bridge.

In 1937, the arched stone pillars over embankments were extended from one to two spans (each side) to accommodate increased street traffic. A similar reconstruction of Krasnoluzhsky Bridge was completed in 1956. The bridge was still in good order when it was demolished to make way for the construction of the Third Ring highway. Space limitations required vertical and horizontal realignment of track (1.5 metres up and 22 metres downstream), so the whole rail bridge (arch, roadway and pillars) had to be rebuilt on a new site (unlike Krasnoluzhsky bridge, which retained its pillars). Demolition (actually, careful disassembly) began in April 1998.

Pushkinsky Pedestrian Bridge (2000)

Instead of scrapping the steel arch of the 1907 Andreyevsky Rail Bridge, city planners re-used it as a structural core for the new pedestrian bridge. May 22, 1999, three barges towed the steel frame to the new anchorage, 1.5 kilometre downstream.

Figure: *An upward view reveals Proskuryakov's original arch*

By this time, contractors already set up concrete foundations, pillars and arches at the new site; they were finished with granite slabs salvaged from the old site. Moskva river at this point is wider than at St.Andrew's, so four pillars and three 25-metre arches over water were required to close the gaps. Pushkinsky bridge connects First Frunzenskaya street in Khamovniki (left bank) with the southern edge of Gorky Park and Titovsky Proezd leading to Leninsky Prospekt (right bank). Design team was led by Yu.P.Platonov. Main pedestrian walkway and stairs are completely enclosed in a glass canopy; there are two open-air side walkways. Left bank entrance has a two-lane escalator. On the right bank, the main flat walkway extends 240 metres past the pillars through the park.

Andreyevsky Rail Bridge (2001)

Initially, rail bridge completion had a higher priority and a 15-month fast track schedule, but in real life priorities changed, and the new road bridge was completed first . Both jobs required removal of old rail bridge, which was done May 22, 1999 after 11 months of preparation . Rail service of the Inner Ring was suspended for two years until completion in 2001. New steel arch (engineer S.S.Tkachenko) is superficially similar to Proskuryakov's original design; it is now 21.5 metre tall, with a higher track alignment; width remains the same, 135.0 metres . The bridge has only one (upstream) pedestrian walkway and two copies of four original obelisks.

Andreyevsky Road Bridge (2000)

New Andreyevsky Road Bridge was completed in December 2000. Design challenge included merging the bridge with the tunnel system under Gagarin Square. Significant incline of roadway ruled out the box girder style in favour of concrete truss arch bridge (lead designer E.G.Gapontsev), set on a 32-metre deep pile foundation . The arch is 135 metres wide and 15 metres high; 39-metre wide roadway has 8 lanes for regular traffic and a single pedestrian and bicycle walkway.

An effective and economical solution, however, failed to preserve visual coherence of two adjacent bridges; new structure completely obstructs the fine outline of Proskuryakov's classic. .

Dallas Water Utilities

Dallas Water Utilities (DWU) is the water and wastewater service operated by the City of Dallas, Texas, in the United States. DWU is a non-profit City of Dallas department that provides services to the city and 31 nearby communities, employs approximately 1450 people, and consists of 26 programs. DWU's budget is completely funded through the rates charged for water and wastewater services provided to customers. Rates are based on the cost of providing the services. The department does not receive any tax revenues. Primary authority and rules for the department are listed in Chapter 49 of the Dallas City Code.

History

Private owners developed the first public water supply in Dallas in 1876. The first sewers were built in the 1880s primarily for storm water drainage. At that time the street runoff and domestic sewage went directly to the Trinity River. Dallas Water Utilities began as the City of Dallas Waterworks in 1881 when the city purchased a privately-owned water company that had been providing Dallas with water since the 1870s. When Browder Springs proved an inadequate water source, the city turned to surface water sources such as the Trinity River and manmade lakes in 1903.

Administration

DWU is a department in the City of Dallas and falls under its management scheme. It is managed by a director and five assistant directors for the five primary functional areas:

Business Operations provides accounting, financial, and budget support to the Water Utilities Department; In addition, this program provides for the management of wholesale water and wastewater services to other governmental entities within the utility's service area provides overall financial support to the Department in areas where the expenditures are not directly tied to the day to day operational and overhead aspects of the department. All such expenditures require special monitoring and control. Key items within General Expense and Debt Service include street rental, transfer to the construction funds, general fund cost reimbursement and debt service commitments

Customer Operations provides customer relations, billings, credit and collections activities, customer information, and all metre-associated services to the water, wastewater and stormwater utility customers. Works to promote DWU conservation programs.

Water Operations is responsible for operating and maintaining the facilities of the Water Utilities Department. These activities include the raw water impoundment and watershed management, and the purification, pumping and distribution of potable water.

Wastewater Operations is responsible for collecting, transporting, controlling the discharge of, and treating domestic and industrial wastes; and maintaining treatment plants and pipelines in the wastewater system.

Capital Improvement Operations plans, designs, constructs and inspects the capital projects needed to provide customers with water and wastewater facilities to meet the growth of the community, extension of water and wastewater mains, modification of facilities to meet changes in State and Federal regulatory requirements (Environmental Protection Agency Administrative Orders, Safe Drinking Water Act treatment parameters and Clean Water Act discharge limitations), and the rehabilitation and replacement of deteriorated or obsolete facilities.

Business Operations

Wholesale Services

DWU has contractual relations with 31 wholesale water and wastewater customers. There are five general types of relationships:

- Wholesale Treated Water Customers
- Wholesale Untreated Water Customers
- Wholesale Wastewater Customers
- Wholesale Untreated Water Customers-Irrigation Only
- Reciprocal Water and Wastewater Customers.

Customer Operations

Conservation Program

Since the early 1980s City of Dallas Water Utilities has had conservation programs. Over the years activities have increased to include children activities, Water-Wise landscape seminars, an annual Water-Wise Landscape tour and more. In 2001, the Dallas City Council took conservation efforts to another level by adopting an irrigation ordinance which included time-of-day watering restrictions.

The ordinance and other measures have reduced gallons per capita per day in Dallas by 21% since 2001 resulting in:

- An estimated water savings of over 98 billion US gallons (370,000,000 m^3) or 12.3 billion US gallons (47,000,000 m^3) annually
- An energy savings of 180 million kilowatt hours of electricity
- A reduction in greenhouse gas emissions by 119,538 tons.

DWU manages the Minor Plumbing repair program which is extremely important in the city's overall commitment to water conservation.

In April 2005, DWU completed a pipeline and delivery system that sent highly treated wastewater from one of its treatment plants directly to the golf links at Cedar Crest Golf Course through the course's irrigation system. During 2005 alone, that new system irrigated Cedar Crest with 81.7 million US gallons (309,000 m^3) of treated wastewater—leaving 81.7 million US gallons (309,000 m^3) of clean water now available for drinking, bathing and other uses.

Metre Reading

DWU has over 300,000 metres in its system. Approximately 6800 are AMI Fixed network units. Metres range in size from 5/8" to 10" or larger.

Water Operations

Water Distribution System

The distribution system of Dallas Water Utilities (DWU) is one of the largest in the United States. As of June 2003, DWU provided retail water service to just over 1.2 million people within the Dallas city limits. The major distribution system facilities include 28 pump stations (including the high service pump stations at the three water treatment plants), 11 ground storage reservoirs, 9 elevated tanks, and 78 vault structures separate from the major facilities, in addition to over 4,600 miles (7,400 km) of distribution and transmission main. The distribution system is divided into 17 pressure zones to maintain adequate water pressures throughout the system. There are four major pressure zones (Central Low, North High, East High, and South High), and five smaller secondary pressure zones (Meandering Way High, Red Bird High, Trinity Heights, Pleasant Grove, and Cedardale) which comprise most of the Dallas service area. Each of the nine major and secondary pressure zones are supplied by one or more pump stations and have elevated or ground storage facilities that establish the static hydraulic gradient for each zone. The remaining eight pressure zones are supplied from an adjacent pressure zone via pressure reducing valve, or single, small booster pump station. These pressure zones do not have storage facilities that establish their static hydraulic gradient.

There are 35 pressure monitoring points throughout the distribution system that are not located at a major facility. These points are connected to the SCADA system, and are used to monitor hydraulic conditions within the distribution system.

The distribution system consists of over 4,600 miles (7,400 km) of pipe with diametres ranging from less than 4 inches (100 mm) to 96 inches (2,400 mm) in diametre. The majority of the total pipe length is sized 8 inches (200 mm) in diametre. About 88 percent of the distribution system is pipes sized 16 inches (410 mm) or smaller in diametre. The DWU distribution system is made of several different pipe materials. Small diametre pipe materials include copper, galvanized iron, PVC, and cast iron. Larger diametre transmission mains are made of steel and various reinforced and prestressed concrete. Approximately 51 percent of the distribution system consists of grey cast iron pipe (CIP). The next most prevalent material types are ductile iron pipe (DIP) and PVC.

Water Sources

Dallas' Long Range Water Supply Plan includes recommendations for water supplies to meet the needs of Dallas and the other cities served through 2050. The plan also includes water conservation and emergency water management plans.

Dallas uses six reservoirs as sources for raw water:

- Lake Ray Hubbard

The City constructed, owns, operates and has available 100% of the permitted water supply for Lake Ray Hubbard. Pumpage of water to nearby treatment facilities began in the summer of 1973. A 1959 water permit applies to the reservoir, its use, and also permits (under limited conditions) storage and usage of water pumped by pipeline from Lake Tawakoni.

- Lewisville Lake
- Grapevine Lake
- Ray Roberts Lake
- Lake Tawakoni
- Lake Fork Reservoir.

Additionally Dallas has water rights to the following sources:

- Lake Palestine—unconnected
- Elm Fork Channel- near Lake Bachman on the Bachman Branch.

The Integrated Pipeline Project is a joint effort between the Tarrant Regional Water District (TRWD) and DWU that will bring additional water supplies to the rapidly growing Dallas/Fort Worth area within the next 10 years. Once completed, this 147-mile (237 km) pipeline will transport water from Lake Palestine, Cedar Creek Reservoir and Richland-Chambers Reservoir back to the TRWD and Dallas service areas. This project will be jointly funded by TRWD and DWU, saving taxpayers millions of dollars and exhibiting a commitment by the region's two largest water providers to work together to meet the region's future water needs. Construction is expected to be complete by 2018.

Water Treatment

The city owns and operates three drinking water treatment plants:

- East Side-The current treatment capacity of the East Side WTP is 440 million US gallons (1,700,000 m^3) per day. An expansion of the East Side WTP to 540 million US gallons (2,000,000 m^3) per day is currently under design, and is projected to be completed by 2012.
- Elm Fork-The current treatment capacity of the Elm Fork WTP is 310 million US gallons (1,200,000 m^3) per day.
- Bachman-The Bachman Water Treatment Plant first began treating water in 1930 at a capacity of 30 million US gallons (110,000 m^3) per day. Since 1930 the plant has undergone several major expansion and minor reservations to bring its capacity to 150 million US gallons (570,000 m^3) per day.

Water Purification

Water purification is the process of removing undesirable chemicals, materials, and biological contaminants from contaminated water. The goal is to produce water fit for a specific purpose. Most water is purified for human consumption (drinking water) but water purification may also be designed for a variety of other purposes, including meeting the requirements of medical, pharmacology, chemical and industrial applications. In general the methods used include physical processes such as filtration and sedimentation, biological processes such as slow sand filters or activated sludge, chemical processes such as flocculation and chlorination and the use of electromagnetic radiation such as ultraviolet light.

The purification process of water may reduce the concentration of particulate matter including suspended particles, parasites, bacteria, algae, viruses, fungi; and a range of dissolved and particulate material derived from the surfaces that water may have made contact with after falling as rain.

The standards for drinking water quality are typically set by governments or by international standards. These standards will typically set minimum and maximum concentrations of contaminants for the use that is to be made of the water. It is not possible to tell whether water is of an appropriate quality by visual examination. Simple procedures such as boiling or the use of a household activated carbon filter are not sufficient for treating all the possible contaminants that may be present in water from an unknown source. Even natural spring water – considered safe for all practical purposes in the 19th century – must now be tested before determining what kind of treatment, if any, is needed. Chemical analysis, while expensive, is the only way to obtain the information necessary for deciding on the appropriate method of purification.

According to a 2007 World Health Organisation report, 1.1 billion people lack access to an improved drinking water supply, 88% of the 4 billion annual cases of diarrheal disease are attributed to unsafe water and inadequate sanitation and hygiene, and 1.8 million people die from diarrheal diseases each year. The WHO estimates that 94% of these diarrheal cases are preventable through modifications to the environment, including access to safe water. Simple techniques for treating water at home, such as chlorination, filters, and solar disinfection, and storing it in safe containers could save a huge number of lives each year. Reducing deaths from waterborne diseases is a major public health goal in developing countries.

Sources of Water

1. Groundwater: The water emerging from some deep ground water may have fallen as rain many tens, hundreds, thousands of years ago. Soil and rock layers naturally filter the ground water to a high degree of clarity before the treatment plant. Such water may emerge as springs, artesian springs, or may be extracted from boreholes or wells. Deep ground water is generally of very high bacteriological quality (i.e., pathogenic bacteria or the pathogenic protozoa are typically absent), but the water typically is rich in dissolved solids, especially carbonates and sulfates of calcium and magnesium. Depending on the strata through which the water has flowed, other ions may also be present including chloride, and bicarbonate. There may be a requirement to reduce the iron or manganese content

of this water to make it pleasant for drinking, cooking, and laundry use. Disinfection may also be required. Where groundwater recharge is practised (a process in which river water is injected into an aquifer to store the water in times of plenty so that it is available in times of drought), the groundwater is equivalent to lowland surface waters for treatment purposes.

2. Upland lakes and reservoirs: Typically located in the headwaters of river systems, upland reservoirs are usually sited above any human habitation and may be surrounded by a protective zone to restrict the opportunities for contamination. Bacteria and pathogen levels are usually low, but some bacteria, protozoa or algae will be present. Where uplands are forested or peaty, humic acids can colour the water. Many upland sources have low pH which require adjustment.
3. Rivers, canals and low land reservoirs: Low land surface waters will have a significant bacterial load and may also contain algae, suspended solids and a variety of dissolved constituents.
4. Atmospheric water generation is a new technology that can provide high quality drinking water by extracting water from the air by cooling the air and thus condensing water vapour.
5. Rainwater harvesting or fog collection which collects water from the atmosphere can be used especially in areas with significant dry seasons and in areas which experience fog even when there is little rain.
6. Desalination of seawater by distillation or reverse osmosis.

Treatment

The processes below are the ones commonly used in water purification plants. Some or most may not be used depending on the scale of the plant and quality of the water.

Pre-treatment

1. Pumping and containment – The majority of water must be pumped from its source or directed into pipes or holding tanks. To avoid adding contaminants to the water, this physical infrastructure must be made from appropriate materials and constructed so that accidental contamination does not occur.
2. Screening – The first step in purifying surface water is to remove large debris such as sticks, leaves, rubbish and other

large particles which may interfere with subsequent purification steps. Most deep groundwater does not need screening before other purification steps.

3. Storage – Water from rivers may also be stored in bankside reservoirs for periods between a few days and many months to allow natural biological purification to take place. This is especially important if treatment is by slow sand filters. Storage reservoirs also provide a buffer against short periods of drought or to allow water supply to be maintained during transitory pollution incidents in the source river.
4. Pre-conditioning – Water rich in hardness salts is treated with soda-ash (sodium carbonate) to precipitate calcium carbonate out utilising the common-ion effect.
5. Pre-chlorination – In many plants the incoming water was chlorinated to minimise the growth of fouling organisms on the pipe-work and tanks. Because of the potential adverse quality effects, this has largely been discontinued.

Widely varied techniques are available to remove the fine solids, micro-organisms and some dissolved inorganic and organic materials. The choice of method will depend on the quality of the water being treated, the cost of the treatment process and the quality standards expected of the processed water.

pH Adjustment

Distilled water has a pH of 7 (neither alkaline nor acidic) and sea water has an average pH of 8.3 (slightly alkaline). If the water is acidic (lower than 7), lime, soda ash, or sodium hydroxide is added to raise the pH. For somewhat acidic waters (lower than 6.5), forced draft degasifiers are the cheapest way to raise the pH, as the process raises the pH by stripping dissolved carbon dioxide (carbonic acid) from the water. Lime is commonly used for pH adjustment for municipal water, or at the start of a treatment plant for process water, as it is cheap, but it also increases the ionic load by raising the water hardness. Making the water slightly alkaline ensures that coagulation and flocculation processes work effectively and also helps to minimise the risk of lead being dissolved from lead pipes and lead solder in pipe fittings. Acid (HCl or H_2SO_4) may be added to alkaline waters in some circumstances to lower the pH. Having alkaline water does not necessarily mean that lead or copper from the plumbing system will not be dissolved into the water but as a generality, water with a pH above 7 is much less likely to dissolve heavy metals than water with a pH below 7.

Figure: *floc floating at the surface of a basin*

Flocculation

Flocculation is a process which clarifies the water. Clarifying means removing any turbidity or colour so that the water is clear and colourless. Clarification is done by causing a precipitate to form in the water which can be removed using simple physical methods. Initially the precipitate forms as very small particles but as the water is gently stirred, these particles stick together to form bigger particles. Many of the small particles that were originally present in the raw water adsorb onto the surface of these small precipitate particles and so get incorporated into the larger particles that coagulation produces. In this way the coagulated precipitate takes most of the suspended matter out of the water and is then filtered off, generally by passing the mixture through a coarse sand filter or sometimes through a mixture of sand and granulated anthracite (high carbon and low volatiles coal). Coagulants / flocculating agents that may be used include:

1. Iron (III) hydroxide. This is formed by adding a solution of an iron (III) compound such as iron(III) chloride to pre-treated water with a pH of 7 or greater. Iron (III) hydroxide is extremely insoluble and forms even at a pH as low as 7. Commercial formulations of iron salts were traditionally marketed in the UK under the name Cuprus.
2. Aluminium hydroxide is also widely used as the flocculating precipitate although there have been concerns about possible health impacts and mishandling led to a severe poisoning incident in 1988 at Camelford in south-west UK when the coagulant was introduced directly into the holding reservoir of final treated water.
3. PolyDADMAC is an artificially produced polymer and is one of a class of synthetic polymers that are now widely used. These polymers have a high molecular weight and form very

stable and readily removed flocs, but tend to be more expensive in use compared to inorganic materials. The materials can also be biodegradable.

Sedimentation

Waters exiting the flocculation basin may enter the sedimentation basin, also called a clarifier or settling basin. It is a large tank with slow flow, allowing floc to settle to the bottom. The sedimentation basin is best located close to the flocculation basin so the transit between does not permit settlement or floc break up. Sedimentation basins may be rectangular, where water flows from end to end, or circular where flow is from the centre outward. Sedimentation basin outflow is typically over a weir so only a thin top layer—that furthest from the sediment—exits. The amount of floc that settles out of the water is dependent on basin retention time and on basin depth. The retention time of the water must therefore be balanced against the cost of a larger basin. The minimum clarifier retention time is normally 4 hours. A deep basin will allow more floc to settle out than a shallow basin. This is because large particles settle faster than smaller ones, so large particles collide with and integrate smaller particles as they settle. In effect, large particles sweep vertically through the basin and clean out smaller particles on their way to the bottom.

As particles settle to the bottom of the basin, a layer of sludge is formed on the floor of the tank. This layer of sludge must be removed and treated. The amount of sludge that is generated is significant, often 3 to 5 percent of the total volume of water that is treated. The cost of treating and disposing of the sludge can be a significant part of the operating cost of a water treatment plant. The tank may be equipped with mechanical cleaning devices that continually clean the bottom of the tank or the tank can be taken out of service when the bottom needs to be cleaned.

Filtration

After separating most floc, the water is filtered as the final step to remove remaining suspended particles and unsettled floc.

Rapid Sand Filters

The most common type of filter is a rapid sand filter. Water moves vertically through sand which often has a layer of activated carbon or anthracite coal above the sand. The top layer removes organic compounds, which contribute to taste and odour. The space between sand particles is larger than the smallest suspended particles, so simple filtration is not enough. Most particles pass through surface layers but are trapped in pore spaces or adhere to sand particles. Effective filtration extends into the depth of the filter. This property of the filter is key to its operation: if the top layer of sand were to block all the particles, the filter would quickly clog.

To clean the filter, water is passed quickly upward through the filter, opposite the normal direction (called *backflushing* or *backwashing*) to remove embedded particles. Prior to this, compressed air may be blown up through the bottom of the filter to break up the compacted filter media to aid the backwashing process; this is known as *air scouring*. This contaminated water can be disposed of, along with the sludge from the sedimentation basin, or it can be recycled by mixing with the raw water entering the plant although this is often considered poor practice since it re-introduces an elevated concentration of bacteria into the raw water

Some water treatment plants employ pressure filters. These work on the same principle as rapid gravity filters, differing in that the filter medium is enclosed in a steel vessel and the water is forced through it under pressure.

Advantages:

- Filters out much smaller particles than paper and sand filters can.
- Filters out virtually all particles larger than their specified pore sizes.
- They are quite thin and so liquids flow through them fairly rapidly.
- They are reasonably strong and so can withstand pressure differences across them of typically 2–5 atmospheres.
- They can be cleaned (back flushed) and reused.

Membrane Filtration

Membrane filters are widely used for filtering both drinking water and sewage. For drinking water, membrane filters can remove virtually all particles larger than 0.2 um—including *giardia* and *cryptosporidium*. Membrane filters are an effective form of tertiary treatment when it is desired to reuse the water for industry, for limited domestic purposes, or before discharging the water into a river that is used by towns further downstream. They are widely used in industry, particularly for beverage preparation (including bottled water). However no filtration can remove substances that are actually dissolved in the water such as phosphorus, nitrates and heavy metal ions.

Slow Sand Filters

Slow sand filters may be used where there is sufficient land and space as the water must be passed very slowly through the filters. These filters rely on biological treatment processes for their action rather than physical filtration. The filters are carefully constructed using graded layers of sand with the coarsest sand, along with some gravel, at the bottom and finest sand at the top. Drains at the base convey treated water away for disinfection. Filtration depends on the development of a thin biological layer, called the zoogleal layer or Schmutzdecke, on the surface of the filter. An effective slow sand filter may remain in service for many weeks or even months if the pre-treatment is well designed and produces water with a very low available nutrient level which physical methods of treatment rarely achieve. Very low nutrient levels allow water to be safely sent through distribution system with very low disinfectant levels thereby reducing consumer irritation over offensive levels of chlorine and chlorine by-products. Slow sand filters are not backwashed; they are maintained by having the top layer of sand scraped off when flow is eventually obstructed by biological growth.

A specific 'large-scale' form of slow sand filter is the process of bank filtration, in which natural sediments in a riverbank are used to provide a first stage of contaminant filtration. While typically not clean enough to be used directly for drinking water, the water gained from the associated extraction wells is much less problematic than river water taken directly from the major streams where bank filtration is often used.

Removal of Ions and Other Dissolved Substances

Ultrafiltration membranes use polymer membranes with chemically formed microscopic pores that can be used to filter out dissolved substances avoiding the use of coagulants. The type of membrane media determines how much pressure is needed to drive the water through and what sizes of micro-organisms can be filtered out.

Ion exchange: Ion exchange systems use ion exchange resin- or zeolite-packed columns to replace unwanted ions. The most common case is water softening consisting of removal of Ca^{2+} and Mg^{2+} ions replacing them with benign (soap friendly) Na^+ or K^+ ions. Ion exchange resins are also used to remove toxic ions such as nitrate, nitrite, lead, mercury, arsenic and many others.

Electrodeionization: Water is passed between a positive electrode and a negative electrode. Ion exchange membranes allow only positive ions to migrate from the treated water toward the negative electrode and only negative ions toward the positive electrode. High purity deionized water is produced with a little worse degree of purification in comparison with ion exchange treatment. Complete removal of ions from water is regarded as electrodialysis. The water is often pre-treated with a reverse osmosis unit to remove non-ionic organic contaminants.

Other Mechanical and Biological Techniques

In addition to the many techniques used in large-scale water treatment, several small-scale, less (or non)-polluting techniques are also being used to treat polluted water. These techniques include those based on mechanical and biological processes. An overview:

- mechanical systems: sand filtration, lava filter systems and systems based on UV-radiation)
- biological systems:
- plant systems as constructed wetlands and treatment ponds (sometimes incorrectly called reedbeds and living walls) and
- compact systems as activated sludge systems, biorotors, aerobic biofilters and anaerobic biofilters, submerged aerated filters, and biorolls.

In order to purify the water adequately, several of these systems are usually combined to work as a whole. Combination of the systems is done in two to three stages, namely primary and secondary purification. Sometimes tertiary purification is also added.

Disinfection

Disinfection is accomplished both by filtering out harmful microbes and also by adding disinfectant chemicals in the last step in purifying drinking water. Water is disinfected to kill any pathogens which pass through the filters. Possible pathogens include viruses, bacteria, including *Escherichia coli*, *Campylobacter* and *Shigella*, and protozoa, including *Giardia lamblia* and other *cryptosporidia*. In most developed countries, public water supplies are required to maintain a residual disinfecting agent throughout the distribution system, in which water may remain for days before reaching the consumer. Following the introduction of any chemical disinfecting agent, the water is usually held in temporary storage – often called a contact tank or clear well to allow the disinfecting action to complete.

Chlorine Disinfection

The most common disinfection method involves some form of chlorine or its compounds such as chloramine or chlorine dioxide. Chlorine is a strong oxidant that rapidly kills many harmful micro-organisms. Because chlorine is a toxic gas, there is a danger of a release associated with its use. This problem is avoided by the use of sodium hypochlorite, which is a relatively inexpensive solution that releases free chlorine when dissolved in water. Chlorine solutions can be generated on site by electrolyzing common salt solutions. A solid form, calcium hypochlorite exists that releases chlorine on contact with water. Handling the solid, however, requires greater routine human contact through opening bags and pouring than the use of gas cylinders or bleach which are more easily automated. The generation of liquid sodium hypochlorite is both inexpensive and safer than the use of gas or solid chlorine. All forms of chlorine are widely used despite their respective drawbacks. One drawback is that chlorine from any source reacts with natural organic compounds in the water to form potentially harmful chemical by-products trihalomethanes (THMs) and haloacetic acids (HAAs), both of which are carcinogenic in large quantities and regulated by the United States Environmental Protection Agency (EPA) and the Drinking Water Inspectorate in the UK. The formation of THMs and haloacetic acids may be minimised by effective removal of as many organics from the water as possible prior to chlorine addition. Although chlorine is effective in killing bacteria, it has limited effectiveness against protozoa that form cysts in water (*Giardia lamblia* and *Cryptosporidium*, both of which are pathogenic).

Chlorine Dioxide Disinfection

Chlorine dioxide is a faster-acting disinfectant than elemental chlorine, however it is relatively rarely used, because in some circumstances it may create excessive amounts of chlorite, which is a by-product regulated to low allowable levels in the United States. Chlorine dioxide is supplied as an aqueous solution and added to water to avoid gas handling problems; chlorine dioxide gas accumulations may spontaneously detonate.

Chloramine Disinfection

The use of chloramine is becoming more common as a disinfectant. Although chloramine is not as strong an oxidant, it does provide a longer-lasting residual than free chlorine and it won't form THMs or haloacetic acids. It is possible to convert chlorine to chloramine by adding ammonia to the water after addition of chlorine. The chlorine and ammonia react to form chloramine. Water distribution systems disinfected with chloramines may experience nitrification, as ammonia is a nutrient for bacterial growth, with nitrates being generated as a by-product.

Ozone Disinfection

O_3 is an unstable molecule which readily gives up one atom of oxygen providing a powerful oxidizing agent which is toxic to most waterborne organisms. It is a very strong, broad spectrum disinfectant that is widely used in Europe. It is an effective method to inactivate harmful protozoa that form cysts. It also works well against almost all other pathogens. Ozone is made by passing oxygen through ultraviolet light or a "cold" electrical discharge. To use ozone as a disinfectant, it must be created on-site and added to the water by bubble contact. Some of the advantages of ozone include the production of fewer dangerous by-products (in comparison to chlorination) and the lack of taste and odour produced by ozonisation. Although fewer by-products are formed by ozonation, it has been discovered that the use of ozone produces a small amount of the suspected carcinogen bromate, although little bromine should be present in treated water. Another of the main disadvantages of ozone is that it leaves no disinfectant residual in the water. Ozone has been used in drinking water plants since 1906 where the first industrial ozonation plant was built in Nice, France. The U.S. Food and Drug Administration has accepted ozone as being safe; and it is applied as an anti-microbiological agent for the treatment, storage, and processing of foods.

Ultraviolet Disinfection

Ultraviolet light is very effective at inactivating cysts, in low turbidity water. UV light's disinfection effectiveness decreases as turbidity increases, a result of the absorption, scattering, and shadowing caused by the suspended solids. The main disadvantage to the use of UV radiation is that, like ozone treatment, it leaves no residual disinfectant in the water; therefore, it is sometimes necessary to add a residual disinfectant after the primary disinfection process. This is often done through the addition of chloramines, discussed above as a primary disinfectant. When used in this manner, chloramines provide an effective residual disinfectant with very few of the negative aspects of chlorination.

Hydrogen Peroxide Disinfection

Works in a similar way to ozone. Activators such as formic acid are often added to increase the efficacy of disinfection. It has the disadvantages that it is slow-working, phytotoxic in high dosage, and decreases the pH of the water it purifies.

Various Portable Methods of Disinfection

Available for disinfection in emergencies or in remote locations. Disinfection is the primary goal, since aesthetic considerations such as taste, odor, appearance, and trace chemical contamination do not affect the short-term safety of drinking water.

Solar Water Disinfection

One low-cost method of disinfecting water that can often be implemented with locally available materials is solar disinfection (SODIS). Unlike methods that rely on firewood, it has low impact on the environment.

One recent study has found that the wild Salmonella which would reproduce quickly during subsequent dark storage of solar-disinfected water could be controlled by the addition of just 10 parts per million of hydrogen peroxide.

Additional Treatment Options

1. Water fluoridation: in many areas fluoride is added to water with the goal of preventing tooth decay. Fluoride is usually added after the disinfection process. In the U.S., fluoridation is usually accomplished by the addition of hexafluorosilicic acid, which decomposes in water, yielding fluoride ions.
2. Water conditioning: This is a method of reducing the effects of hard water. Hardness salts are deposited in water systems subject to heating because the decomposition of bicarbonate ions creates carbonate ions that crystallise out of the saturated solution of calcium or magnesium carbonate. Water with high concentrations of hardness salts can be treated with soda ash (sodium carbonate) which precipitates out the excess salts, through the common-ion effect, producing calcium carbonate of very high purity. The precipitated calcium carbonate is

traditionally sold to the manufacturers of toothpaste. Several other methods of industrial and residential water treatment are claimed (without general scientific acceptance) to include the use of magnetic or/and electrical fields reducing the effects of hard water.

3. Plumbo Solvency Reduction: In areas with naturally acidic waters of low conductivity (i.e. surface rainfall in upland mountains of igneous rocks), the water may be capable of dissolving lead from any lead pipes that it is carried in. The addition of small quantities of phosphate ion and increasing the pH slightly both assist in greatly reducing plumbo-solvency by creating insoluble lead salts on the inner surfaces of the pipes.
4. Radium Removal: Some groundwater sources contain radium, a radioactive chemical element. Typical sources include many groundwater sources north of the Illinois River in Illinois. Radium can be removed by ion exchange, or by water conditioning. The back flush or sludge that is produced is, however, a low-level radioactive waste.
5. Fluoride Removal: Although fluoride is added to water in many areas, some areas of the world have excessive levels of natural fluoride in the source water. Excessive levels can be toxic or cause undesirable cosmetic effects such as staining of teeth. Methods of reducing fluoride levels is through treatment with activated alumina and bone char filter media.

Other Water Purification Techniques

Other popular methods for purifying water, especially for local private supplies are listed below. In some countries some of these methods are also used for large scale municipal supplies. Particularly important are distillation (de-salination of seawater) and reverse osmosis.

1. Boiling: Water is heated hot enough and long enough to inactivate or kill micro-organisms that normally live in water at room temperature. Near sea level, a vigorous rolling boil for at least one minute is sufficient. At high altitudes (greater than two kilometres or 5000 feet) three minutes is recommended. In areas where the water is "hard" (that is, containing significant dissolved calcium salts), boiling decomposes the bicarbonate ions, resulting in partial precipitation as calcium carbonate. This is the "fur" that builds up on kettle elements, etc., in hard water areas. With the exception of calcium, boiling does not

remove solutes of higher boiling point than water and in fact increases their concentration (due to some water being lost as vapour). Boiling does not leave a residual disinfectant in the water. Therefore, water that has been boiled and then stored for any length of time may have acquired new pathogens.

2. Granular Activated Carbon filtering: a form of activated carbon with a high surface area, adsorbs many compounds including many toxic compounds. Water passing through activated carbon is commonly used in municipal regions with organic contamination, taste or odors. Many household water filters and fish tanks use activated carbon filters to further purify the water. Household filters for drinking water sometimes contain silver as metallic silver nanoparticle. If water is held in the carbon block for longer period, microorganisms can grow inside which results in fouling and contamination. Silver nanoparticles are excellent anti-bacterial material and they can decompose toxic halo-organic compounds such as pesticides into non-toxic organic products.
3. Distillation involves boiling the water to produce water vapour. The vapour contacts a cool surface where it condenses as a liquid. Because the solutes are not normally vaporised, they remain in the boiling solution. Even distillation does not completely purify water, because of contaminants with similar boiling points and droplets of unvapourised liquid carried with the steam. However, 99.9% pure water can be obtained by distillation.
4. Reverse osmosis: Mechanical pressure is applied to an impure solution to force pure water through a semi-permeable membrane. Reverse osmosis is theoretically the most thorough method of large scale water purification available, although perfect semi-permeable membranes are difficult to create. Unless membranes are well-maintained, algae and other life forms can colonise the membranes.
5. The use of iron in removing arsenic from water.
6. Direct contact membrane distillation (DCMD). Applicable to desalination. Heated seawater is passed along the surface of a hydrophobic polymer membrane. Evaporated water passes from the hot side through pores in the membrane into a stream of cold pure water on the other side. The difference in vapour pressure between the hot and cold side helps to push water molecules through.

7. Gas hydrate crystals centrifuge method. If carbon dioxide gas is mixed with contaminated water at high pressure and low temperature, gas hydrate crystals will contain only clean water. This is because the water molecules bind to the gas molecules at molecular level. The contaminated water is in liquid form. A centrifuge may be used to separate the crystals and the concentrated contaminated water.
8. In Situ Chemical Oxidation , a form of advanced oxidation processes and advanced oxidation technology, is an environmental remediation technique used for soil and/or groundwater remediation to reduce the concentrations of targeted environmental contaminants to acceptable levels. ISCO is accomplished by injecting or otherwise introducing strong chemical oxidizers directly into the contaminated medium (soil or groundwater) to destroy chemical contaminants in place. It can be used to remediate a variety of organic compounds, including some that are resistant to natural degradation.

Hydrogen Production

For the small scale production of hydrogen, water purifiers are installed to prevent formation of minerals on the surface of the electrodes and to remove organics and chlorine from utility water. First, the water passes through a 20 micrometre interference (mesh or screen filter) filter to remove sand and dust particles, then a charcoal filter using activated carbon to remove organics and chlorine and finally a de-ionizing filter to remove metallic ions. Testing can be done before and after the filter to verify the proper removal of barium, calcium, potassium, magnesium, sodium and silica.

Another method that is used is reverse osmosis.

Safety and Controversies

In April, 2007, the water supply of Spencer, Massachusetts became contaminated with excess sodium hydroxide (lye) when its treatment equipment malfunctioned.

Many municipalities have moved from free chlorine to chloramine as a disinfection agent. However, chloramine in some water systems, appears to be a corrosive agent. Chloramine can dissolve the "protective" film inside older service line, with the leaching of lead into residential spigots. This can result in harmful exposure to lead, with elevated blood levels of lead the outcome. Lead is a known neurotoxin.

Demineralised Water

Distillation removes all minerals from water, and the membrane methods of reverse osmosis and nanofiltration remove most to all minerals. This results in demineralised water which is not considered ideal drinking water. The World Health Organisation has investigated the health effects of demineralised water since 1980. Experiments in humans found that demineralised water increased diuresis and the elimination of electrolytes, with decreased blood serum potassium concentration. Magnesium, calcium, and other minerals in water can help to protect against nutritional deficiency. Demineralised water may also increase the risk from toxic metals because it more readily leaches materials from piping like lead and cadmium, which is prevented by dissolved minerals such as calcium and magnesium. Low-mineral water has been implicated in specific cases of lead poisoning in infants, when lead from pipes leached at especially high rates into the water. Recommendations for magnesium have been put at a minimum of 10 mg/L with 20–30 mg/L optimum; for calcium a 20 mg/L minimum and a 40–80 mg/L optimum, and a total water hardness (adding magnesium and calcium) of 2 to 4 mmol/L. At water hardness above 5 mmol/L, higher incidence of gallstones, kidney stones, urinary stones, arthrosis, and arthropathies have been observed. Additionally, desalination processes can increase the risk of bacterial contamination.

Manufacturers of home water distillers claim the opposite—that minerals in water are the cause of many diseases, and that most beneficial minerals come from food, not water. They quote the American Medical Association as saying "The body's need for minerals is largely met through foods, not drinking water." The WHO report agrees that "drinking water, with some rare exceptions, is not the major source of essential elements for humans" and is "not the major source of our calcium and magnesium intake", yet states that demineralised water is harmful anyway. "Additional evidence comes from animal experiments and clinical observations in several countries. Animals given zinc or magnesium dosed in their drinking water had a significantly higher concentration of these elements in the serum than animals given the same elements in much higher amounts with food and provided with low-mineral water to drink."

5

Agricultural Drainage System

An agricultural drainage system is a system by which the water level on or in the soil is controlled to enhance agricultural crop production.

Classification

The function of the field drainage system is to control the water table, whereas the function of the main drainage system is to collect, transport, and dispose of the water through an outfall or outlet. In some instances one makes an additional distinction between collector and main drainage systems. Field drainage systems are differentiated in surface and subsurface field drainage systems.

Sometimes (e.g. in irrigated, submerged rice fields), a form of temporary drainage is required whereby the drainage system is allowed to function on certain occasions only (e.g. during the harvest period). If allowed to function continuously, excessive quantities of water would be lost. Such a system is therefore called a checked, or controlled, drainage system. More usually, however, the drainage system is meant to function as regularly as possible to prevent undue waterlogging at any time and one employs a regular drainage system. In literature, this is sometimes also called a "relief drainage system".

Surface Drainage Systems

The regular surface drainage systems, which start functioning as soon as there is an excess of rainfall or irrigation, operate entirely by gravity. They consist of reshaped or reformed land surfaces and can be divided into:

- Bedded systems, used in flat lands for crops other than rice;
- Graded systems, used in sloping land for crops other than rice.

The bedded and graded systems may have ridges and furrows.

The checked surface drainage systems consist of check gates placed in the embankments surrounding flat basins, such as those used for rice fields in flat lands. These fields are usually submerged and only need to be drained on certain occasions (e.g. at harvest time). Checked surface drainage systems are also found in terraced lands used for rice.

In literature, not much information can be found on the relations between the various regular surface field drainage systems, the reduction in the degree of waterlogging, and the agricultural or environmental effects. It is therefore difficult to develop sound agricultural criteria for the regular surface field drainage systems. Most of the known criteria for these systems concern the efficiency of the techniques of land levelling and earthmoving. Similarly, agricultural criteria for checked surface drainage systems are not very well known.

Subsurface Drainage Systems

Like the surface field drainage systems, the subsurface field drainage systems can also be differentiated in regular systems and checked (controlled) systems.

When the drain discharge takes place entirely by gravity, both types of subsurface systems have much in common, except that the *checked* systems have control gates that can be opened and closed according to need. They can save much irrigation water. A *checked* drainage system also reduces the discharge through the main drainage system, thereby reducing construction costs.

When the discharge takes place by *pumping*, the drainage can be checked simply by not operating the pumps or by reducing the pumping time. In northwestern India, this practice has increased the irrigation efficiency and reduced the quantity of irrigation water needed, and has not led to any undue salinization.

The subsurface field drainage systems consist of horizontal or slightly sloping channels made in the soil; they can be open ditches, trenches, filled with brushwood and a soil cap, filled with stones and a soil cap, buried pipe drains, tile drains, or mole drains, but they can also consist of a series of wells.

Modern buried pipe drains often consist of corrugated, flexible, and perforated plastic (PE or PVC) pipe lines wrapped with an *envelope* or filter material to improve the permeability around the pipes and to prevent entry of soil particles, which is especially important in fine sandy and silty soils. The surround may consist of synthetic fibre (geotextile).

The *field drains* (or *laterals*) discharge their water into the collector or main system either by *gravity* or by *pumping*.

The wells (which may be open dug wells or *tubewells*) have normally to be pumped, but sometimes they are connected to drains for discharge by gravity.

Subsurface drainage by wells is often referred to as vertical drainage, and drainage by channels as horizontal drainage, but it is more clear to speak of "field drainage by wells" and "field drainage by ditches or pipes" respectively.

In some instances, subsurface drainage can be achieved simply by breaking up slowly permeable soil layers by *deep plowing* (*sub-soiling*), provided that the underground has sufficient natural drainage. In other instances, a combination of sub-soiling and subsurface drains may solve the problem.

Main Drainage Systems

Deep collector drains are required for subsurface field drainage systems, whereas shallow collector drains are used for surface field drainage systems, but they can also be used for pumped subsurface systems. The deep collectors may consist of open ditches or buried pipe lines.

The terms *deep collectors* and *shallow collectors* refer rather to the depth of the water level in the collector below the soil surface than to the depth of the bottom of the collector. The bottom depth is determined both by the depth of the water level and by the required discharge capacity.

The deep collectors may either discharge their water into deep main drains (which are drains that do not receive water directly from field drains, but only from collectors), or their water may be pumped into a disposal drain.

Disposal drains are main drains in which the depth of the water level below the soil surface is not bound to a minimum, and the water level may even be above the soil surface, provided that embankments are made to prevent inundation. Disposal drains can serve both subsurface and surface field drainage systems.

Deep main drains can gradually become disposal drains if they are given a smaller gradient than the land slope along the drain.

The technical criteria applicable to main drainage systems depend on the hydrological situation and on the type of system.

Main Drainage Outlet

The final point of a main drainage system is the gravity outlet structure or the pumping station.

Applications

Surface drainage systems are usually applied in relatively flat lands that have soils with a low or medium infiltration capacity, or in lands with high-intensity rainfalls that exceed the normal infiltration capacity, so that frequent waterlogging occurs on the soil surface.

Subsurface drainage systems are used when the drainage problem is mainly that of shallow water tables. When both surface and subsurface waterlogging occur, a combined surface/subsurface drainage system is required. Sometimes, a subsurface drainage system is installed in soils with a low infiltration capacity, where a surface drainage problem may improve the soil structure and the infiltration capacity so greatly that a surface drainage system is no longer required.

On the other hand, it can also happen that a surface drainage system diminishes the recharge of the groundwater to such an extent that the subsurface drainage problem is considerably reduced or even eliminated.

The choice between a subsurface drainage system by pipes and ditches or by tube wells is more a matter of technical criteria and costs than of agricultural criteria, because both types of systems can be designed to meet the same agricultural criteria and achieve the same benefits. Usually, pipe drains or ditches are preferable to wells. However, when the soil consists of a poorly permeable top layer several metres thick, overlying a rapidly permeable and deep subsoil, wells may be a better option, because the drain spacing required for pipes or ditches would be considerably smaller than the spacing for wells.

When the land needs a subsurface drainage system, but saline groundwater is present at great depth, it is better to employ a shallow, closely spaced system of pipes or ditches instead of a deep, widely spaced system. The reason is that the deeper systems produce a more salty effluent than the shallow systems. Environmental criteria may then prohibit the use of the deeper systems.

In some drainage projects, one may find that only main drainage systems are envisaged. The agricultural land is then still likely to suffer from field drainage problems. In other cases, one may find that field drainage systems are ineffective because there is no adequate main drainage system. In either case, the installation of drainage systems is not recommended. Reference: gives a general description of land drainage in the world and shows a paper on types of agricultural land drainage systems used in different parts of the world.

Drainage System Design

The analysis of positive and negative (side) effects of drainage and the optimisation of drainage design in accordance to the *drainage design procedures* is discussed in the article on Drainage research.

Mine Reclamation

Mine reclamation is the process of creating useful landscapes that meet a variety of goals, typically creating productive ecosystems (or sometimes industrial or municipal land) from mined land. It includes all aspects of this work, including material placement, stabilising, capping, regrading, placing cover soils, revegetation, and maintenance.

Modern mine rehabilitation aims to minimise and mitigate the environmental effects of modern mining, which may in the case of open pit mining involve movement of significant volumes of rock. Rehabilitation management is an ongoing process, often resulting in open pit mines being backfilled.

After mining finishes, the mine area must undergo rehabilitation.

- Waste dumps are contoured to flatten them out, to further stabilise them against erosion.
- If the ore contains sulfides it is usually covered with a layer of clay to prevent access of rain and oxygen from the air, which can oxidize the sulfides to produce sulfuric acid.
- Landfills are covered with topsoil, and vegetation is planted to help consolidate the material.
- Dumps are usually fenced off to prevent livestock denuding them of vegetation.
- The open pit is then surrounded with a fence, to prevent access, and it generally eventually fills up with groundwater.

- Tailings dams are left to evaporate, then covered with waste rock, clay if need be, and soil, which is planted to stabilise it.

For underground mines, rehabilitation is not always a significant problem or cost. This is because of the higher grade of the ore and lower volumes of waste rock and tailings. In some situations, stopes are backfilled with concrete slurry using waste, so that minimal waste is left at surface.

The removal of plant and infrastructure is not always part of a rehabilitation program, as many old mine plants have cultural heritage and cultural value. Often in gold mines, rehabilitation is performed by scavenger operations which treat the soil within the plant area for spilled gold using modified placer mining gravity collection plants.

In the United States, mine reclamation is a regular part of modern mining practice.

Stream Restoration

Stream restoration or river restoration, sometimes called river reclamation in the UK, describes a set of activities that help improve the environmental health of a river or stream. Improved health may be indicated by expanded habitat for diverse species (e.g. fish, aquatic insects, other wildlife) and reduced stream bank erosion. Enhancements may also include improved water quality (i.e. reduction of pollutant levels and increased dissolved oxygen levels) and achieving a self-sustaining, functional flow regime in the stream system that does not require periodic human intervention, such as dredging or construction of flood control structures. Stream restoration projects can also yield increased property values in adjacent areas.

Stream restoration differs from:

- river engineering, a term which typically refers to alteration of a water body for a non-environmental benefit such as navigation, flood control or water supply diversion;
- waterway restoration, a term used in the United Kingdom describing alterations to a canal or river to improve navigability and related recreational amenities.

Restoration Techniques

Restoration activities may range from a simple removal of a disturbance which inhibits natural stream function (e.g. repairing a damaged culvert), to stabilisation of stream banks, to more active intervention such as installation of stormwater management facilities, such as riparian zone restoration and constructed wetlands.

Figure: *Robinson Creek restoration project (2005) included re-shaping of stream bank slopes, addition of live willow and large rock baffles, removal of invasive species and revegetation with indigenous species.*

Successful restoration projects begin with careful study of the stream system, including the historical weather patterns, stream hydraulics, sediment transport patterns and related conditions. Researchers evaluating restoration projects have found that many of these projects subsequently fail (e.g., with flooding or excessive erosion) because the projects were not designed with a sufficient scientific basis; restoration techniques may have been selected for aesthetic reasons.

In-stream Techniques

Channel Modification

Modifications to a stream channel may be appropriate to address degradation. Channel modifications may yield improved habitat for wildlife and plants in a stream corridor, but can result in flooding, excessive erosion or other damage if not carefully planned. Design of modifications involves a careful analysis of a complex fluvial processes. Alterations may include channel shape (in terms of sinuosity and meander characteristics), cross-section and channel profile (slope along the channel bed). Alterations affect the dissipation of energy through the channel, which has an impact on stream velocity and turbulence, sediment volume and size distribution, scour, and water surface elevations, among other characteristics.

Cross-vanes and Related Structures

A cross-vane is a "U"-shaped structure of boulders or logs, built across the channel to reduce velocity and energy near the stream banks. It reduces bank erosion, maintains channel capacity and provides other benefits such as improved habitat for aquatic species. Similar structures used to dissipate stream energy include the W-Weir and J-Hook Vane.

Engineered Log Jams

An emerging stream restoration technique is the installation of *engineered log jams.* Reintroduction of large woody debris into a stream is a fairly recent method that is being experimented with in streams such as Thornton Creek, in Seattle, WA. Because of channelization and removal of woody debris, many streams now lack the hydraulic complexity that is necessary to maintain bank stabilisation and a healthy plant/animal habitat. Engineered log jams are individually designed to meet the needs of specific restoration projects, but there are overarching design elements. One element is to anchor logs along the stream bank in order to create a physical blockade against erosion. A second element of engineered log jams is to improve fish habitat. Log jams add diversity to the water flow by creating riffles, pools, and temperature variations. This is vital to fish because it provides the right circumstances to spawn, rest, feed, hunt, and hide. There are inherent dangers to engineered log dams. If not properly implemented, they cause erosion and sediment in unwanted areas leading to more damage than repair.

Off-line Techniques

As part of a stream restoration project, stormwater management facilities may be installed in the immediate corridor or in upland areas. These facilities, which reduce the velocity and/or the volume of stormwater entering the stream channel, can also improve water quality, and include:

- Bioretention systems and rain gardens
- Constructed wetlands
- Infiltration basins
- Retention basins.

Monitoring of Restoration Projects

Sponsors of restoration projects may conduct monitoring of stream conditions after construction, to evaluate effectiveness. In some projects it may take considerable time before there is evidence of desired biological activity, such as fish spawning, therefore monitoring efforts may be conducted for several years after a restoration project has completed.

Water Reclamation

Water reclamation is a process by which wastewater from homes and businesses is cleaned using biological and chemical treatment so that the water can be returned to the environment safely to augment the natural systems from which it came. It is used today as both an aquifer and stream enhancement strategy.

Benefits

Water reclamation helps decrease diverging water from sensitive eco-systems which depend greatly on the flow to improve the quality of the water. Water reclamation also decreases the pollution to bodies of water, such as oceans and rivers, by diverting the wastewater. Some of the pollutants are used for irrigation purposes, such as nitrogen, which would be harmful to these bodies of water. Another benefit is the enhancement of wetlands which benefits the wildlife dependent on that eco-system. For instance, The San Jose/Santa Clara Water Pollution Control Plant instituted a recycling program to protect the San Francisco Bay area's natural salt water marshes.

Uses

Most of the uses of water reclamation are non potable uses such as: Washing Cars, flushing toilets, cooling water for power plants, concrete mixing, artificial lakes, and irrigation for golf courses and public parks. Most systems run a dual piping system to keep the recycled water separate from the potable water. Indirect water reclamation (water passing through a body of water first) will result in potable water usage. While, this is not practiced in the United States, it is in other countries such as Namibia.

Aboard the International Space Station, astronauts have been able to drink recycled urine due to the introduction of the ECLSS system. The system costs $250 million and has been working since May 2009. The system recycles wastewater and urine back into potable water used for drinking, food preparation, and oxygen generation. This cuts back on the need for resupplying the space station as often.

Alternatives to Water Reclamation

Greywater

Like water reclamation, greywater recycles used water for irrigation and other various uses. Greywater uses the same waste as water reclamation with the exception of toilet water. Unlike water reclamation, greywater does not purify the water through a biological process, but by sand filters and soil. After passing through the soil, where it is filtered like rain water, it flows into the groundwater where it will be reused again.

Desalination

Desalination is an energy-intensive process where salt and other minerals are removed from sea water to produce potable water for drinking and irrigation, typically through membrane filtration (reverse-osmosis), and steam-distillation.

Many ships and submarines make use of this technology. The idea of desalination is a heavily researched area worldwide. Many methodologies have been developed, including nuclear-powered desalination, and solar-powered desalination.

Reclamation Processes

Wastewater must pass through numerous systems before being returned to the environment. Here is a partial listing from one particular plant system:

- Barscreens - Barscreens remove large solids that are sent into a grinder. All solids are then dumped into a sewer pipe at a Treatment Plant.
- Primary Settling Tanks - Readily settable and floatable solids are removed from the wastewater. These solids are skimmed from the top and bottom of the tanks and sent to the Treatment Plant where it'll be turned into fertilizer.
- Biological Treatment - The wastewater is cleaned through a biological treatment method that uses microorganisms, bacteria which digest the sludge and reduce the nutrient content. Air bubbles up to keep the organisms suspended and to supply oxygen to the aerobic bacteria so they can metabolize the food, convert it to energy, CO_2, and water, and reproduce more microorganisms. This helps to remove ammonia also through nitrification.
- Secondary Settling Tanks - The force of the flow slows down as sewage enters these tanks, allowing the microorganisms to settle to the bottom. As they settle, other small particles suspended in the water are picked up, leaving behind clear wastewater. Some of the microorganisms that settle to the bottom are returned to the system to be used again.

- Tertiary Treatment - Deep-bed, single-media, gravity sand filters receive water from the secondary basins and filter out the remaining solids. As this is the final process to remove solids, the water in these filters is almost completely clear.
- Chlorine Contact Tanks - Three chlorine contact tanks disinfect the water to decrease the risks associated with discharging wastewater containing human pathogens. This step protects the quality of the waters that receive the wastewater discharge.

One of two procedures are then followed according to the future disposal site:

1. Reclaimed Water Pump Station - The pump station distributes reclaimed water to users around the City. This may include golf courses, agricultural uses, cooling towers, or in land fills.
2. Water is passed through high level purification to be returned to the environment. Currently this means a reverse osmosis system.

Reclamation of Wellington Harbour

The reclamation of Wellington Harbour started in the 1850s, originally to increase the amount of usable flat land for Wellington city. Reclamations in the 1960s and 1970s were to meet the needs of container shipping (containerisation) and new cargo handling methods. Reclamation has added more than 155 hectares to Wellington.

The Location of Wellington

A plan for the New Zealand Company's new settlement of Britannia at Pito-one (Petone) had been prepared in England by Samuel Cobham. The key elements of his city were a large amount of flat land on the shores of a harbour, traversable by a navigable river. When surveyors arrived in 1840 on the *Cuba* led by Captain William Mein Smith, it was determined that the Hutt River was not navigable and, due to its tendency to flood, was not appropriate to support a major city. For these reasons the new settlement was relocated to the southern shores of Port Nicholson and renamed Wellington.

Edward Gibbon Wakefield of the New Zealand Company had devised a system of 'packages' of land for colonists of one town acre each. Cobham's Brittania consited of 1100 1-acre (4,000 m^2) town sections, which William Mein Smith struggled to fit into the new location. These sections were squeezed into the available space by sacrificing many of the planned amenities such as parks, reserves, ports, libraries and many other public areas identified in the original plan. For this reason, from Wellington's outset, there was a need for extra land.

19th Century Reclamations and the Establishment of the Wellington Harbour Board

While large scale reclamation began in the 1850s, the earliest reclamations in Wellington were conducted by private citizens. A popular story of the first reclamation conducted in Wellington was that done by George Bennet. Bennet had arrived in 1848 on the *Berenicia* and purchased a hilly section at Windy, or Clay Point (what is now the corner of Lambton Quay and Willis Street). At that time, Windy Point was a precipice with a narrow and often impassible path connecting Willis Street to Beach Road (now Lambton Quay). Bennet commenced, to the amusement of neighbours, with pick-axe, shovel and barrow to move earth from the Point, tossing the spoil onto the path and into the harbour, thus widened the track and performing Wellington's first reclamation.

Figure: *Two men fishing off the reclamations at Te Aro, in the vicinity of Cable Street, Wellington. Photograph taken circa. 1910.*

A programme of systematic reclamation began in 1852, overseen by the provincial government. Charles Rooking Carter completed a 360' x 100' extension below Willis Street at a cost of £1,036.

In 1855, the magnitude 8.2 Wairarapa Earthquake uplifted the northwestern side of Wellington bay (in some places up to 1.5 metres). This created a tidal swamp, and rendered many of the existing jetties in the harbour unusable. Most of this land was subsequently reclaimed, providing an excellent new rail and road route to the north. Another result of the newly-raised land in Wellington was that the shipping basin planned for the city was abandoned and the land used for a cricket ground, the Basin Reserve.

The Wellington City Council was inaugurated in 1870, and by the end of the 1870s some 70 acres (280,000 m^2) of land had been reclaimed using spoil from the hills behind Lambton Quay and from Wadestown Hill. In 1880, the Wellington Harbour Board was established to manage and develop the harbour and its facilities. From then reclamation work was divided between the Harbour Board, the Government and the City Council. Among major developments from 1880 to the turn of the century was the reclamation north of Pipitea Point for railways land and south of Queens Wharf to Te Aro by the City Council. This removed the last vestiges of private ownership of the foreshore, putting the waterfront under the control of the Harbour Board. By the end of the 19th century, the original 1840 shoreline was unrecognisable.

Twentieth Century Reclamations

From 1900 to 1930 further reclamations were made for railways and Harbour Board purposes. Additional wharves and the seawall at Oriental Bay were built, and the boat harbour at Clyde Quay was constructed.

1960s and Container Shipping

The final phase of reclamation took place in the 1960s and 1970s. A government report in 1967 recommended the adoption of containerisation and that Wellington should be one of the two New Zealand ports. With containerisation came new roll-on/roll-off cargo handling methods that require more land adjacent to ships' berths. This resulted in the start of an extension to the Aotea Quay reclamation. Reclamation was carried out on both side of Queens Wharf and the Wellington Harbour Board Container Terminal was created by a large reclamation at Thorndon.

The first container ship berthed on 19 June 1971. The container terminal has 24.3 hectares of back-up space capable of holding 6,284 containers.

Recent History and Present Day

After the advent of the Port Reform Act in 1988, the Wellington Harbour Board, along with all other harbour boards, ceased to exist after October 1989. Its commercial, property management and recreational functions were split between the Port of Wellington, Lambton Harbour Management and Wellington City Council respectively.

In 1976, the Historic Places Trust placed 14 plaques along the original shoreline. These plaques run from Pipitea Point, along Lambton Quay, through Mercer Street, lower Cuba Street, Wakefield Street to Oriental Parade at the northern corner of Herd Street.

Reclamation by Year, Location, and Area

Table: *From the New Zealand Electronic Text Centre*

Year/s	*Location*	*Area*
1852	Willis Street, Mercer Street, Chew's Lane, Bonds	
1857-63	Bank of N.Z., corner Willis Street and Lambton Quay (Noah's Ark site), Harris Street to Grey Street	7 acres ($28,000\ m^2$)
1859	Oddfellows' Hall site	
1864	Foresters' Lodge site	
1865	Messrs. Jacob Joseph, between Waring Taylor and Stout Streets and Lambton Quay	total of last three, 2 rods
1866–67	Panama, Brandon, Johnston and Waring Taylor Streets, pts. Featherston Street and Customhouse Quay	12 acres ($49,000\ m^2$)
1875	Government Building site	2 acres ($8,100\ m^2$)
1876	Government (Lambton) railway station and lines, Featherston Street extension, Ballance, Stout, Bunny and Whitmore Streets, Govt. Printing Office and "Shacks" (this was extended to Pipitea Point)	46 acres ($190,000\ m^2$)
1882	Manawatu (Thorndon) railway station and lines	29 acres ($120,000\ m^2$)
1882	Railway Wharf	1 rod
1884	Davis Street Extension	
1886	Hunter Street endowment, Customhouse Quay and Hunter Street	3 rods
1886	Victoria and Wakefield Streets	22 acres ($89,000\ m^2$)
1889	Jervois Quay	17 acres ($69,000\ m^2$)
1893	Harbour Board store, and track for Te Aro railway, Customhouse Quay and Jervois Quay	1-acre ($4,000\ m^2$)
1893–1901	Waterloo Quay and Glasgow wharves	3 acres ($12,000\ m^2$)

Contd...

Year/s	***Location***	***Area***
1895	Council's yards, near Oriental Parade	1-acre (4,000 m^2)
1901–03	From near Queen's Wharf to the Lyttelton Ferry Wharf, and site of Customhouse	2.5 acres (10,000 m^2)
1901–1914	Barnet, Cable and Chaffers Street	18 acres (73,000 m^2)
1902–1925	Clyde Quay widening	4 acres (16,000 m^2)
1904	Hutt Road locality	
1904–1916	Waterloo and near Fryatt Quays, Hinemoa and Cornwall Streets	34 acres (140,000 m^2)
1906	Waterloo Quay completion	34.5 acres (140,000 m^2)
1906	Oriental Parade and boat shed sites	1-acre (4,000 m^2)
1910–1913	Davis Street extension, near Cornwall Street	4 acres (16,000 m^2)
1924–1927	Thorndon Esplanade and Hutt railway lines areas, vested in the Harbour Board and Railway Department	

Seismic Retrofitting

Seismic retrofitting is the modification of existing structures to make them more resistant to seismic activity, ground motion, or soil failure due to earthquakes. With better understanding of seismic demand on structures and with our recent experiences with large earthquakes near urban centres, the need of seismic retrofitting is well acknowledged. Prior to the introduction of modern seismic codes in the late 1960s for developed countries (US, Japan etc.) and late 1970s for many other parts of the world (Turkey, China etc.), many structures were designed without adequate detailing and reinforcement for seismic protection. In view of the imminent problem, various research work has been carried out. Furthermore, state-of-the-art technical guidelines for seismic assessment, retrofit and rehabilitation have been published around the world - such as the ASCE-SEI 41 and the New Zealand Society for Earthquake Engineering (NZSEE)'s guidelines.

The retrofit techniques outlined here are also applicable for other natural hazards such as tropical cyclones, tornadoes, and severe winds from thunderstorms. Whilst current practice of seismic retrofitting is predominantly concerned with structural improvements to reduce the seismic hazard of using the structures, it is similarly essential to reduce the hazards and losses from non-structural elements. It is also important to keep in mind that there is no such thing as an earthquake-proof structure, although seismic performance can be greatly enhanced through proper initial design or subsequent modifications.

Strategies

Seismic retrofit (or rehabilitation) strategies have been developed in the past few decades following the introduction of new seismic provisions and the availability of advanced materials (e.g. fibre-reinforced polymers, FRP, fibre reinforced concrete and high strength steel). Retrofit strategies are different from retrofit techniques, where the former is the basic approach to achieve an overall retrofit performance objective, such as increasing strength, increasing deformability, reducing deformation demands while the latter is the technical methods to achieve that strategy, for example FRP jacketing.

- Increasing the global capacity (strengthening). This is typically done by the addition of cross braces or new structural walls.
- Reduction of the seismic demand by means of supplementary damping and/or use of base isolation systems.
- Increasing the local capacity of structural elements. This strategy recognises the inherent capacity within the existing structures, and therefore adopt a more cost-effective approach to selectively upgrade local capacity (deformation/ductility, strength or stiffness) of individual structural components.
- Selective weakening retrofit. This is a counter intuitive strategy to change the inelastic mechanism of the structure, whilst recognising the inherent capacity of the structure.
- Allowing sliding connections such as passageway bridges to accommodate additional movement between seismically independent structures.

Performance Objectives

In the past, seismic retrofit was primarily applied to achieve public safety, with engineering solutions limited by economic and political considerations. However, with the development of Performance based earthquake engineering (PBEE), several levels of performance objectives are gradually recognised:

- Public safety only. The goal is to protect human life, ensuring that the structure will not collapse upon its occupants or passersby, and that the structure can be safely exited. Under severe seismic conditions the structure may be a total economic write-off, requiring tear-down and replacement.
- Structure survivability. The goal is that the structure, while remaining safe for exit, may require extensive repair (but not replacement) before it is generally useful or considered safe for occupation. This is typically the lowest level of retrofit applied to bridges.
- Structure functionality. Primary structure undamaged and the structure is undiminished in utility for its primary application. A high level of retrofit, this ensures that any required repairs are only "cosmetic" - for example, minor cracks in plaster, drywall and stucco. This is the minimum acceptable level of retrofit for hospitals.
- Structure unaffected. This level of retrofit is preferred for historic structures of high cultural significance.

Techniques

Common seismic retrofitting techniques fall into several categories:

External Post-tensioning

The use of external post-tensioning for new structural systems have been developed in the past decade. Under the PRESS (Precast Seismic Structural Systems), a large-scale U.S./Japan joint research program, unbonded post-tensioning high strength steel tendons have been used to achieve a moment-resisting system that has self-centring capacity. An extension of the same idea for seismic retrofitting has been experimentally tested for seismic retrofit of California bridges under a Caltrans research project and for seismic retrofit of non-ductile reinforced concrete frames. Pre-stressing can increase the capacity of structural elements such as beam, column and beam-column joints. It should be noted that external pre-stressing has been used for structural upgrade for gravity/live loading since 1970s

Base Isolators

Base isolation is a collection of structural elements of a building that should substantially decouple the building's structure from the shaking ground thus protecting the building's integrity and enhancing its seismic performance. This earthquake engineering technology, which is a kind of seismic vibration control, can be applied both to a newly designed building and to seismic upgrading of existing structures. Normally, excavations are made around the building and the building is separated from the foundations. Steel or reinforced concrete beams replace the connections to the foundations, while under these, the isolating pads, or base isolators, replace the material removed. While the base isolation tends to restrict transmission of the ground motion to the building, it also keeps the building positioned properly over the foundation. Careful attention to detail is required where the building interfaces with the ground, especially at entrances, stairways and ramps, to ensure sufficient relative motion of those structural elements.

Supplementary Dampers

Supplementary dampers absorb the energy of motion and convert it to heat, thus "damping" resonant effects in structures that are rigidly attached to the ground. In addition to adding energy dissipation capacity to the structure, supplementary damping can reduce the displacement and acceleration demand within the structures. In some cases, the threat of damage does not come from the initial shock itself, but rather from the periodic resonant motion of the structure that repeated ground motion induces. In partical sense, supplementary dampers act similarly to Shock absorbers used in automotive suspensions.

Tuned Mass Dampers

Tuned mass dampers (TMD) employ movable weights on some sort of springs. These are typically employed to reduce wind sway in very tall, light buildings. Similar designs may be employed to impart earthquake resistance in eight to ten story buildings that are prone to destructive earthquake induced resonances.

Slosh Tank

A slosh tank is a large tank of fluid placed on an upper floor. During a seismic event, the fluid in this tank will slosh back and forth, but is directed by baffles - partitions that prevent the tank itself becoming resonant; through its mass the water may change or counter the resonant period of the building. Additional kinetic energy can be converted to heat by the baffles and is dissipated through the water - any temperature rise will be insignificant.

Active Control System

Very tall buildings ("skyscrapers"), when built using modern lightweight materials, might sway uncomfortably (but not dangerously) in certain wind conditions. A solution to this problem is to include at some upper story a large mass, constrained, but free to move within a limited range, and moving on some sort of bearing system such as an air cushion or hydraulic film. Hydraulic pistons, powered by electric pumps and accumulators, are actively driven to counter the wind forces and natural resonances.

These may also, if properly designed, be effective in controlling excessive motion - with or without applied power - in an earthquake. In general, though, modern steel frame high rise buildings are not as subject to dangerous motion as are medium rise (eight to ten story) buildings, as the resonant period of a tall and massive building is longer than the approximately one second shocks applied by an earthquake.

Adhoc Addition of Structural Support/Reinforcement

The most common form of seismic retrofit to lower buildings is adding strength to the existing structure to resist seismic forces. The strengthening may be limited to connections between existing building elements or it may involve adding primary resisting elements such as walls or frames, particularly in the lower stories.

Connections between Buildings and their Expansion Additions

Frequently, building additions will not be strongly connected to the existing structure, but simply placed adjacent to it, with only minor continuity in flooring, siding, and roofing. As a result, the addition may have a different resonant period than the original structure, and they may easily detach from one another.

The relative motion will then cause the two parts to collide, causing severe structural damage. Proper construction will tie the two building components rigidly together so that they behave as a single mass or employ dampers to expend the energy from relative motion, with appropriate allowance for this motion.

Exterior Reinforcement of Building

Exterior Concrete Columns: Historic buildings, made of unreinforced masonry, may have culturally important interior detailing or murals that should not be disturbed. In this case, the solution may be to add a number of steel, reinforced concrete, or poststressed concrete columns to the exterior. Careful attention must be paid to the connections with other members such as footings, top plates, and roof trusses.

Infill Shear Trusses

In this case, there was sufficient vertical strength in the building columns and sufficient shear strength in the lower stories that only limited shear reinforcement was required to make it earthquake resistant for this location near the Hayward fault

Massive Exterior Structure

In other circumstances, far greater reinforcement is required. A parking garage over shops — the placement, detailing, and painting of the reinforcement becomes itself an architectural embellishment.

Typical Retrofit Scenario and Solution

Soft-story Failure

This collapse mode is known as *soft story collapse.* In many buildings the ground level is designed for different uses than the upper levels. Low rise residential structures may be built over a parking garage which have large doors on one side. Hotels may have a tall ground floors to allow for a grand entrance or ballrooms. Office buildings may have stores in the ground floor which desire continuous windows for display.

Traditional seismic design assumes that the lower stories of a building are stronger than the upper stories and where this is not the case—if the lower story is less strong than the upper structure—the structure will not respond to earthquakes in the expected fashion. Using modern design methods, it is possible to take a weak story into account. Several failures of this type in one large apartment complex caused most of the fatalities in the 1994 Northridge earthquake.

Typically, where this type of problem is found, the weak story is reinforced to make it stronger than the floors above by adding shear walls or moment frames. Moment frames consisting of inverted Ubents are useful in preserving lower story garage access, while a lower cost solution may be to use shear walls or trusses in several locations, which partially reduce the usefulness for automobile parking but still allow the space to be used for other storage.

Beam-column Joint Connections

Beam-column joint connections are a common structural weakness in dealing with seismic retrofitting. Prior to the introduction of modern seismic codes in early 1970s, beam-column joints were typically non-engineered or designed. Laboratory testings have confirmed the seismic vulnerability of these poorly detailed and under-designed connections. Failure of beam-column joint connections can typically lead to catastrophic collapse of a frame-building, as often observed in recent earthquakes

Figure: *Corner joint steel reinforcement and high tensile strength rods with grouted anti-burst jacket below*

For reinforced concrete beam-column joints - various retrofit solutions have been proposed and tested in the past 20 years. Philosophically, the various seismic retrofit strategies discussed above can be implemented for reinforced concrete joints.

Concrete or steel jacketing have been a popular retrofit technique until the advent of composite materials such as Carbon fibre-reinforced polymer (FRP). Composite materials such as carbon FRP and aramic FRP have been extensively tested for use in seismic retrofit with some success. One novel technique includes the use of selective weakening of the beam and added external post-tensioning to the joint in order to achieve flexural hinging in the beam, which is more desirable in terms of seismic design.

Widespread weld failures at beam-column joints of low-to-medium rise steel buildings during the Northridge 1994 earthquake for example, have shown the structural deficiencies of these 'modern-designed' post-1970s welded moment-resisting connections. A subsequent SAC research project has documented, tested and proposed several retrofit solutions for these welded steel moment-resisting connections. Various retrofit solutions have been developed for these welded joints - such as a) weld strengthening and b) addition of steel haunch or 'dog-bone' shape flange.

Shear Failure Within Floor Diaphragm

Floors in wooden buildings are usually constructed upon relatively deep spans of wood, called joists, covered with a diagonal wood planking or plywood to form a subfloor upon which the finish floor surface is laid. In many structures these are all aligned in the same direction. To prevent the beams from tipping over onto their side, blocking is used at each end, and for additional stiffness, blocking or diagonal wood or metal bracing may be placed between beams at one or more points in their spans. At the outer edge it is typical to use a single depth of blocking and a perimetre beam overall.

If the blocking or nailing is inadequate, each beam can be laid flat by the shear forces applied to the building. In this position they lack most of their original strength and the structure may further collapse. As part of a retrofit the blocking may be doubled, especially at the outer edges of the building. It may be appropriate to add additional nails between the sill plate of the perimetre wall erected upon the floor diaphragm, although this will require exposing the sill plate by removing interior plaster or exterior siding. As the sill plate may be quite old and dry and substantial nails must be used, it may be necessary to pre-drill a hole for the nail in the old wood to avoid splitting. When the wall is opened for this purpose it may also be appropriate to tie vertical wall elements into the foundation using speciality connectors and bolts glued with epoxy cement into holes drilled in the foundation.

Sliding Off Foundation and "Cripple Wall" Failure

Single or two story wood-frame domestic structures built on a perimetre or slab foundation are relatively safe in an earthquake, but in many structures built before 1950 the sill plate that sits between the concrete foundation and the floor diaphragm (perimetre foundation) or studwall (slab foundation) may not be sufficiently bolted in. Additionally, older attachments (without substantial corrosion-proofing) may have corroded to a point of weakness. A sideways shock can slide the building entirely off of the foundations or slab.

Often such buildings, especially if constructed on a moderate slope, are erected on a platform connected to a perimetre foundation through low stud-walls called "cripple wall" or *pin-up*. This low wall structure itself may fail in shear or in its connections to itself at the corners, leading to the building moving diagonally and collapsing the low walls. The likelihood of failure of the pin-up can be reduced by ensuring that the corners are well reinforced in shear and that the shear panels are well connected to each other through the corner posts.

Figure: *House slid off of foundation*

This requires structural grade sheet plywood, often treated for rot resistance. This grade of plywood is made without interior unfilled knots and with more, thinner layers than common plywood. New buildings designed to resist earthquakes will typically use OSB (oriented strand board), sometimes with metal joins between panels, and with well attached stucco covering to enhance its performance. In many modern tract homes, especially those built upon expansive (clay) soil the building is constructed upon a single and relatively thick monolithic slab, kept in one piece by high tensile rods that are stressed after the slab has set.

This poststressing places the concrete under compression - a condition under which it is extremely strong in bending and so will not crack under adverse soil conditions!

Multiple Piers in Shallow Pits

Some older low-cost structures are elevated on tapered concrete pylons set into shallow pits, a method frequently used to attach outdoor decks to existing buildings. This is seen in conditions of damp soil, especially in tropical conditions, as it leaves a dry ventilated space under the house, and in far northern conditions of permafrost (frozen mud) as it keeps the building's warmth from destabilising the ground beneath. During an earthquake, the pylons may tip, spilling the building to the ground. This can be overcome by using deep-bored holes to contain cast-in-place reinforced pylons, which are then secured to the floor panel at the corners of the building. Another technique is to add sufficient diagonal bracing or sections of concrete shear wall between pylons.

Reinforced Concrete Column Burst

Reinforced concrete columns typically contain large diametre vertical rebar (reinforcing bars) arranged in a ring, surrounded by lighter-gauge hoops of rebar. Upon analysis of failures due to earthquakes, it has been realised that the weakness was not in the vertical bars, but rather in inadequate strength and quantity of hoops. Once the integrity of the hoops is breached, the vertical rebar can flex outward, stressing the central column of concrete.

Figure: *Jacketed and grouted column on left, unmodified on right*

The concrete then simply crumbles into small pieces, now unconstrained by the surrounding rebar. In new construction a greater amount of hoop-like structures are used. One simple retrofit is to surround the column with a jacket of steel plates formed and welded into a single cylinder. The space between the jacket and the column is then filled with concrete, a process called grouting. Where soil or structure conditions require such additional modification, additional pilings may be driven near the column base and concrete pads linking the pilings to the pylon are fabricated at or below ground level. In the example shown not all columns needed to be modified to gain sufficient seismic resistance for the conditions expected. (This location is about a mile from the Hayward Fault Zone.)

Reinforced Concrete Wall Burst

Concrete walls are often used at the transition between elevated road fill and overpass structures. The wall is used both to retain the soil and so enable the use of a shorter span and also to transfer the weight of the span directly downward to footings in undisturbed soil. If these walls are inadequate they may crumble under the stress of an earthquake's induced ground motion.

One form of retrofit is to drill numerous holes into the surface of the wall, and secure short L-shaped sections of rebar to the surface of each hole with epoxy adhesive. Additional vertical and horizontal rebar is then secured to the new elements, a form is erected, and an additional layer of concrete is poured. This modification may be combined with additional footings in excavated trenches and additional support ledgers and tie-backs to retain the span on the bounding walls.

Brick Wall Resin and Glass Fibre Reinforcement

Brick building structures have been reinforced with coatings of glass fibre and appropriate resin (epoxy or polyester). In lower floors these may be applied over entire exposed surfaces, while in upper floors this may be confined to narrow areas around window and door openings. This application provides tensile strength that stiffens the wall against bending away from the side with the application. The efficient protection of an entire building requires extensive analysis and engineering to determine the appropriate locations to be treated.

Lift

Where moist or poorly consolidated alluvial soil interfaces in a "beach like" structure against underlying firm material, seismic waves travelling through the alluvium can be amplified, just as are water waves against a sloping beach. In these special conditions, vertical accelerations up to twice the force of gravity have been measured. If a building is not secured to a well-embedded foundation it is possible for the building to be thrust from (or with) its foundations into the air, usually with severe damage upon landing. Even if it is well-founded, higher portions such as upper stories or roof structures or attached structures such as canopies and porches may become detached from the primary structure.

Good practices in modern, earthquake-resistant structures dictate that there be good vertical connections throughout every component of the building, from undisturbed or engineered earth to foundation to sill plate to vertical studs to plate cap through each floor and continuing to the roof structure.

Above the foundation and sill plate the connections are typically made using steel strap or sheet stampings, nailed to wood members using special hardened high-shear strength nails, and heavy angle stampings secured with through bolts, using large washers to prevent pull-through. Where inadequate bolts are provided between the sill plates and a foundation in existing construction (or are not trusted due to possible corrosion), special clamp plates may be added, each of which is secured to the foundation using expansion bolts inserted into holes drilled in an exposed face of concrete. Other members must then be secured to the sill plates with additional fittings.

Soil

One of the most difficult retrofits is that required to prevent damage due to soil failure. Soil failure can occur on a slope, a slope failure or landslide, or in a flat area due to liquefaction of water-saturated sand and/or mud. Generally, deep pilings must be driven into stable soil (typically hard mud or sand) or to underlying bedrock or the slope must be stabilised.

For buildings built atop previous landslides the practicality of retrofit may be limited by economic factors, as it is not practical to stabilise a large, deep landslide. The likelihood of landslide or soil failure may also depend upon seasonal factors, as the soil may be more stable at the beginning of a wet season than at the beginning of the dry season. Such a "two season" *Mediterranean climate* is seen throughout California.

In some cases, the best that can be done is to reduce the entrance of water runoff from higher, stable elevations by capturing and bypassing through channels or pipes, and to drain water infiltrated directly and from subsurface springs by inserting horizontal perforated tubes. There are numerous locations in California where extensive developments have been built atop archaic landslides, which have not moved in historic times but which (if both water-saturated and shaken by an earthquake) have a high probability of moving *en masse*, carrying entire sections of suburban development to new locations. While the most modern of house structures (well tied to monolithic concrete foundation slabs reinforced with post tensioning cables) may survive such movement largely intact, the building will no longer be in its proper location.

Utility Pipes and Cables: Risks

Natural gas and propane supply pipes to structures often prove especially dangerous during and after earthquakes. Should a building move from its foundation or fall due to cripple wall collapse, the ductile iron pipes transporting the gas within the structure may be broken, typically at the location of threaded joints. The gas may then still be provided to the pressure regulator from higher pressure lines and so continue to flow in substantial quantities; it may then be ignited by a nearby source such as a lit pilot light or arcing electrical connection.

There are two primary methods of automatically restraining the flow of gas after an earthquake, installed on the low pressure side of the regulator, and usually downstream of the gas metre.

- A caged metal ball may be arranged at the edge of an orifice. Upon seismic shock, the ball will roll into the orifice, sealing it to prevent gas flow. The ball may later be reset by the use of an external magnet. This device will respond only to ground motion.
- A flow-sensitive device may be used to close a valve if the flow of gas exceeds a set threshold (very much like an electrical circuit breaker). This device will operate independently of seismic motion, but will not respond to minor leaks which may be caused by an earthquake.

It appears that the most secure configuration would be to use one of each of these devices in series.

Tunnels

Unless the tunnel penetrates a fault likely to slip, the greatest danger to tunnels is a landslide blocking an entrance. Additional protection around the entrance may be applied to divert any falling material (similar as is done to divert snowavalanches) or the slope above the tunnel may be stabilised in some way. Where only small- to medium-sized rocks and boulders are expected to fall, the entire slope may be covered with wire mesh, pinned down to the slope with metal rods. This is also a common modification to highway cuts where appropriate conditions exist.

Underwater Tubes

The safety of underwater tubes is highly dependent upon the soil conditions through which the tunnel was constructed, the materials and reinforcements used, and the maximum predicted earthquake expected, and other factors, some of which may remain unknown under current knowledge.

BART Tube

A tube of particular structural, seismic, economic, and political interest is the BART (Bay Area Rapid Transit) trans-bay tube. This tube was constructed at the bottom of San Francisco Bay through an innovative process. Rather than pushing a shield through the soft bay mud, the tube was constructed on land in sections. Each section consisted of two inner train tunnels of circular cross section, a central access tunnel of rectangular cross section, and an outer oval shell encompassing the three inner tubes. The intervening space was filled with concrete. At the bottom of the bay a trench was excavated and a flat bed of crushed stone prepared to receive the tube sections. The sections were then floated into place and sunk, then joined with bolted connections to previously-placed sections. An overfill was then placed atop the tube to hold it down. Once completed from San Francisco to Oakland, the tracks and electrical components were installed. The predicted response of the tube during a major earthquake was likened to be as that of a string of (cooked) spaghetti in a bowl of gelatin dessert. To avoid overstressing the tube due to differential movements at each end, a slidingslip joint was included at the San Francisco terminus under the landmark Ferry Building.

The engineers of the construction consortium PBTB (Parsons Brinckerhoff-Tudor-Bechtel) used the best estimates of ground motion available at the time, now known to be insufficient given modern computational analysis methods and geotechnical knowledge. Unexpected settlement of the tube has reduced the amount of slip that can be accommodated without failure. These factors have resulted in the slip joint being designed too short to ensure survival of the tube under possible (perhaps even likely) large earthquakes in the region. To correct this deficiency the slip joint must be extended to allow for additional movement, a modification expected to be both expensive and technically and logistically difficult. Other retrofits to the BART tube include vibratory consolidation of the tube's overfill to avoid potential liquefying of the overfill, which has now been completed. (Should the overfill fail there is a danger of portions of the tube rising from the bottom, an event which could potentially cause failure of the section connections.)

Bridge Retrofit

Bridges have several failure modes.

Expansion Rockers

Many short bridge spans are statically anchored at one end and attached to rockers at the other. This rocker gives vertical and transverse support while allowing the bridge span to expand and contract with temperature changes. The change in the length of the span is accommodated over a gap in the roadway by comb-like expansion joints. During severe ground motion, the rockers may jump from their tracks or be moved beyond their design limits, causing the bridge to unship from its resting point and then either become misaligned or fail completely. Motion can be constrained by adding ductile or high-strength steel restraints that are friction-clamped to beams and designed to slide under extreme stress while still limiting the motion relative to the anchorage.

Deck Rigidity

Figure: *Additional diagonals were inserted under both decks of this bridge*

Suspension bridges may respond to earthquakes with a side-to-side motion exceeding that which was designed for wind gust response. Such motion can cause fragmentation of the road surface, damage to bearings, and plastic deformation or breakage of components. Devices such as hydraulic dampers or clamped sliding connections and additional diagonal reenforcement may be added.

Lattice Girders, Beams, and Ties

Lattice girders consist of two "I"-beams connected with a criss-cross lattice of flat strap or angle stock. These can be greatly strengthened by replacing the open lattice with plate members. This is usually done in concert with the replacement of hot rivets with bolts.

Hot Rivets

Many older structures were fabricated by inserting red-hot rivets into pre-drilled holes; the soft rivets are then peened using an air hammer on one side and a bucking bar (an inertial mass) on the head end. As these cool slowly, they are left in an annealed (soft) condition, while the plate, having been hot rolled and quenched during manufacture, remains relatively hard. Under extreme stress the hard plates can shear the soft rivets, resulting in failure of the joint. The solution is to burn out each rivet with an oxygen torch. The hole is then prepared to a precise diametre with a reamer. A special *locator bolt*, consisting of a head, a shaft matching the reamed hole, and a threaded end is inserted and retained with a nut, then tightened with a wrench. As the bolt has been formed from an appropriate high-strength alloy and has also been heat-treated, it is not subject to either the plastic shear failure typical of hot rivets nor the brittle fracture of ordinary bolts. Any partial failure will be in the plastic flow of the metal secured by the bolt; with proper engineering any such failure should be non-catastrophic.

Fill and Overpass

Elevated roadways are typically built on sections of elevated earth fill connected with bridge-like segments, often supported with vertical columns. If the soil fails where a bridge terminates, the bridge may become disconnected from the rest of the roadway and break away. The retrofit for this is to add additional reinforcement to any supporting wall, or to add deep caissons adjacent to the edge at each end and connect them with a supporting beam under the bridge.

Another failure occurs when the fill at each end moves (through resonant effects) in bulk, in opposite directions. If there is an insufficient founding shelf for the overpass, then it may fall. Additional shelf and ductile stays may be added to attach the overpass to the footings at one or both ends. The stays, rather than being fixed to the beams, may instead be clamped to them. Under moderate loading, these keep the overpass centred in the gap so that it is less likely to slide off its founding shelf at one end. The ability for the fixed ends to slide, rather than break, will prevent the complete drop of the structure if it should fail to remain on the footings.

Viaducts

Large sections of roadway may consist entirely of viaduct, sections with no connection to the earth other than through vertical columns. When concrete columns are used, the detailing is critical. Typical failure may be in the toppling of a row of columns due either to soil connection failure or to insufficient cylindrical wrapping with rebar. Both failures were seen in the 1995 Great Hanshin earthquake in Kobe, Japan, where an entire viaduct, centrally supported by a single row of large columns, was laid down to one side. Such columns are reinforced by excavating to the foundation pad, driving additional pilings, and adding a new, larger pad, well connected with rebar alongside or into the column. A column with insufficient wrapping bar, which is prone to burst and then hinge at the bursting point, may be completely encased in a circular or elliptical jacket of welded steel sheet and grouted as described above.

Figure: *Cypress Freeway viaduct collapse. Note failure of inadequate anti-burst wrapping and lack of connection between upper and lower vertical elements.*

Sometimes viaducts may fail in the connections between components. This was seen in the failure of the Cypress Freeway in Oakland, California, during the Loma Prieta earthquake. This viaduct was a two-level structure, and the upper portions of the columns were not well connected to the lower portions that supported the lower level; this caused the upper deck to collapse upon the lower deck. Weak connections such as these require additional external jacketing - either through external steel components or by a complete jacket of reinforced concrete, often using stub connections that are glued (using epoxy adhesive) into numerous drilled holes. These stubs are then connected to additional wrappings, external forms (which may be temporary or permanent) are erected, and additional concrete is poured into the space. Large connected structures similar to the Cypress Viaduct must also be properly analysed in their entirety using dynamic computer simulations.

Wood Frame Structure

Predominantly residential/dwelling in North America consisted of wood-frame structure. Wood is one of the best materials for anti-seismic construction since it is of low mass and is relatively less brittle than masonry. It is easy to work with and very cheap compared to other modern material as steel and reinforced concrete. This is only resistant if the structure is properly connected to its foundation and has adequate shear resistance, in modern construction obtained by well connected surfacing of panels with plywood or oriented strand board in combination with exterior stucco. Steel strapping and sheet forms are also used to connect elements securely.

Retrofit methods in older woodframe structures may consist of the following, and other methods not described here.

- The lowest plate rails of walls are bolted to a continuous foundation, or held down with rigid metal clips bolted to the foundation.
- Selected vertical elements, especially at wall junctures and window and door openings are attached securely to the sill plate.
- In two story buildings using "western" style construction (walls are progressively erected upon the lower story's upper diaphragm, unlike "eastern" *balloon framing*), the upper walls are connected to the lower walls with tension elements. In some cases, connections may be extended vertically to include retention of certain roof elements.
- Low *cripple walls* are made shear resistant by adding plywood at the corners, and by securing corners from opening with metal strapping or fixtures.
- Vertical posts may be restrained from jumping off of their footings.

Wooden framing is efficient when combined with masonry, if the structure is properly designed. In Turkey, the traditional houses (bagdadi) are made with this technology. In El Salvador, wood and bamboo are used for residential construction.

Reinforced and Unreinforced Masonry

In many parts of developing countries such as Pakistan, Iran and China, unreinforced or in some cases reinforced masonry is the predominantly form of structures for rural residential and dwelling. Masonry was also a common construction form in the early part of the 20th century, which implies that a substantial number of these at-risk masonry structures would have significant heritage value. Masonry walls that are not reinforced are especially hazardous. Such structures may be more appropriate for replacement than retrofit, but if the walls are the principal load bearing elements in structures of modest size they may be appropriately reinforced. It is especially important that floor and ceiling beams be securely attached to the walls. Additional vertical supports in the form of steel or reinforced concrete may be added.

In the western United States, much of what is seen as masonry is actually brick or stone veneer. Current construction rules dictate the amount of *tie–back* required, which consist of metal straps secured to vertical structural elements. These straps extend into mortar courses, securing the veneer to the primary structure. Older structures may not secure this sufficiently for seismic safety. A weakly secured veneer in a house interior (sometimes used to face a fireplace from floor to ceiling) can be especially dangerous to occupants. Older masonry chimneys are also dangerous if they have substantial vertical extension above the roof. These are prone to breakage at the roofline and may fall into the house in a single large piece. For retrofit, additional supports may be added or it may be better to simply remove the extension and replace it with lighter materials, with special piping replacing the flue tile and a wood structure replacing the masonry. This may be matched against existing brickwork by using very thin veneer (similar to a tile, but with the appearance of a brick).

Destructive Testing

In destructive testing, tests are carried out to the specimen's failure, in order to understand a specimen's structural performance or material behaviour under different loads.

Destructive testing is most suitable, and economic, for objects which will be mass produced, as the cost of destroying a small number of specimens is negligible. It is usually not economical to do destructive testing where only one or very few items are to be produced (for example, in the case of a building).

Some types of destructive testing:

- Stress tests
- Crash tests
- Hardness tests
- Metallographic tests.

Testing of Large Structures

Building structures or large nonbuilding structures (such as dams and bridges) are rarely subjected to destructive testing due to the prohibitive cost of constructing a building, or a scale model of a building, just to destroy it.

Earthquake engineering requires a good understanding of how structures will perform at earthquakes. Destructive tests are more frequently carried out for structures which are to be constructed in earthquake zones. Such tests are sometimes referred to as *crash tests*, and they are carried out to verify the designed seismic performance of a new building, or the actual performance of an existing building. The tests are, mostly, carried out on a platform called a shake-table which is designed to shake in the same manner as an earthquake. Results of those tests often include the corresponding shake-table videos. Testing of structures in earthquakes is increasingly done by modelling the structure using specialist finite element software.

Destructive Software Testing

Destructive software testing is a type of software testing which attempts to cause a piece of software to fail in an uncontrolled manner, in order to test its robustness. The Earthquake Engineering Research Institute (EERI) is a leading technical society in dissemination of earthquake risk and earthquake engineering research both in the U.S. and globally.

EERI members include researchers, geologists, geotechnical engineers, educators, government officials, and building code regulators. Their mission, as stated in their 5 year plan published in 2006, has three points.

1. "Advancing the science and practice of earthquake engineering"
2. "Improving understanding of the impact of earthquakes on the physical, social, economic, political, and cultural environment"
3. "Advocating comprehensive and realistic measures for reducing the harmful effects of earthquakes"

Goals

In the 2006 5 year plan, the EERI has identified four main goals towards fulfilling their mission and planned strategies to carry them out.

1. "Enhance and expand educational materials and technical programs." They will hold two seminars per year on topics intended to interest a wide audience. They will also post many of their publications online, such as their journal *Earthquake Spectra*.
2. "Outreach and Advocacy" They will continue to release their findings on earthquake risks, including the costs of potential

disasters. They hope to influence policymakers to increase funds for preventing these risks. They hope also to include earthquake safety into the "green" building design movement.

3. "Maintain a strong program of international activities." They serve as an inflow point in the U.S. for earthquake research from other countries. They also serve as an outflow, translating their research into languages other than English.
4. "Expand and broaden financial resource base." They wish to raise $1 million in donations by 2010, and increase worldwide membership to 3,000. They wish to expand their programs and partnerships with other organisations with more workshops and seminars. In February, 2010, the EERI entered a partnership with the Geo- Institute of ASCE, increasing their collaboration to reduce earthquake hazards.

Earthquake Engineering Research Institute

The Earthquake Engineering Research Institute (EERI) is a leading technical society in dissemination of earthquake risk and earthquake engineering research both in the U.S. and globally. EERI members include researchers, geologists, geotechnical engineers, educators, government officials, and building code regulators. Their mission, as stated in their 5 year plan published in 2006, has three points.

1. "Advancing the science and practice of earthquake engineering"
2. "Improving understanding of the impact of earthquakes on the physical, social, economic, political, and cultural environment"
3. "Advocating comprehensive and realistic measures for reducing the harmful effects of earthquakes"

History

History section cites Tubbesing, Susan K., and Thalia Anagnos. "The Earthquake Engineering Research Institute, a Short History of the U.S. National Earthquake Engineering Society."

The EERI was formed in 1948 as an advising committee on the U.S. Coast and Geodetic Survey. It quickly became its own independent, nonprofit organisation, with the purpose of studying why buildings fail under earthquake disasters, and what methods can prevent these failures. At first they conducted their research in laboratories of different University or Government groups. As the EERI grew, they began to more often send research funds to the Universities, and have the University conduct the research. EERI focused more on identifying and investigating areas in need of research, and policymaking based on the university's lab results.

In 1952 the EERI organised the first Conference on Earthquake Engineering, at UCLA. In 1955, they held the first World Conference on Earthquake Engineering. In 1984, the 8th World Conference was held in San Francisco. This conference brought in scientists from 54 countries.

At first, membership to the EERI was limited to invite-only engineers and scientists. In 1973, they began to hire members by application, and increased their membership from 126 to 721 by 1978. In 1991, EERI began receiving funding from the Federal Emergency Management Agency (FEMA), to continue publishing information on how to reduce damage from earthquakes.

After a number of location changes, the EERI headquarters settled in Oakland, California. Their quarterly journal, *Earthquake Spectra*, covers current research on earthquake engineering and is available online or by subscription. Its target audience is any geologist, seismologist, or related engineer. EERI also publishes many other types of information, including a monthly newsletter, an oral history series, and field investigation reports.

California Earthquake Assessments

EERI performs risk assessments on earthquake potential sites around the world. This is a quick summary of two reports on California cities. In 2006 an engineering firm related to the EERI has projected over $122 billion in damages, if a repeat of the 1906 San Fransisco earthquake occurs. This number includes damages to homes and structures, excluding fire damage. The EERI lobbies for government funding to prevent natural disasters. The money is best spent before loss of life and large scale structural damage, though often it is not seen until afterward, as evidenced by Hurricane Katrina. The EERI and the USGS have identified that a potential large earthquake in Los Angeles would cause more damage than Katrina at New Orleans, with up to $250 billion in total damages and 18,000 deaths.

Student Involvement

A few representatives from each chapter make up the Student Leadership Council (SLC). Since 2008 the EERI and SLC have held the Undergraduate Seismic Design Competition, which was previously run by the Pacific Earthquake Engineering Research Centre (PEER). In this competition a team of undergraduate college students must design and construct a structure made of balsa wood. The structure is limited by many rules, such as a weight limit, the individual heights of each floor, total height limit, and more.

The structure is subjected to extra weight and placed on a shake table, which moves to simulate an earthquake. An accelerometre is placed on top of the building to measure how fast the top of the building shakes. Students' structures are judged on a number of criteria, including the height of the structure, number of floors, the accelerometre readings, and whether the structure breaks. Students will want to make a building close to the height limit because the higher floors are worth more points. The 8th annual competition is to be held at San Diego, February 11 and 12, along with the 63rd EERI annual meeting.

6

Water Supply

Water supply is the provision by public utilities, commercial organisations, community endeavours or by individuals of water, usually by a system of pumps and pipes. Irrigation is covered separately.

Figure: *Water from a tap - supplied by a pipe.*

Global Access to Clean Water

In 2010 about 84% of the global population (6.74 billion people) had access to piped water supply through house connections or to an improved water source through other means than house, including standpipes, "water kiosks", protected springs and protected wells. However, about 14% (884 million people) did not have access to an improved water source and had to use unprotected wells or springs, canals, lakes or rivers for their water needs.

A clean water supply, especially so with regard to sewage, is the single most important determinant of public health. Destruction of water supply and/or sewage disposal infrastructure after major catastrophes (earthquakes, floods, war, etc.) poses the immediate threat of severe epidemics of waterborne diseases, several of which can be life-threatening.

Figure: *Shipot, an underground water source in Ukraine*

Technical Overview

Water supply systems get water from a variety of locations, including groundwater (aquifers), surface water (lakes and rivers), conservation and the sea through desalination. The water is then, in most cases, purified, disinfected through chlorination and sometimes fluoridated.

Treated water then either flows by gravity or is pumped to reservoirs, which can be elevated such as water towers or on the ground. Once water is used, wastewater is typically discharged in a sewer system and treated in a wastewater treatment plant before being discharged into a river, lake or the sea or reused for landscaping, irrigation or industrial use.

Service Quality

Many of the 3.5 billion people having access to piped water receive a poor or very poor quality of service, especially in developing countries where about 80% of the world population lives. Water supply service quality has many dimensions: continuity; water quality; pressure; and the degree of responsiveness of service providers to customer complaints.

Continuity of Supply

Continuity of water supply is taken for granted in most developed countries, but is a severe problem in many developing countries, where sometimes water is only provided for a few hours every day or a few days a week. It is estimated that about half of the population of developing countries receives water on an intermittent basis.

Water Quality

Drinking water quality has a micro-biological and a physico-chemical dimension. There are thousands of parameters of water quality. In public water supply systems water should, at a minimum, be disinfected — most commonly through the use of chlorination or the use of ultra violet light — or it may need to undergo treatment, especially in the case of surface water.

Water Pressure

Water pressures vary in different locations of a distribution system. Water mains below the street may operate at higher pressures, with a pressure reducer located at each point where the water enters a building or a house. In poorly managed systems, water pressure can be so low as to result only in a trickle of water or so high that it leads to damage to plumbing fixtures and waste of water. Pressure in an urban water system is typically maintained either by a pressurised water tank serving an urban area, by pumping the water up into a tower and relying on gravity to maintain a constant pressure in the system or solely by pumps at the water treatment plant and repeater pumping stations.

***Figure:** 1880s model of pumping engine, in Herne Bay Museum*

Typical UK pressures are 4-5 bar for an urban supply. However, some people can get over eight bars or below one bar. A single iron main pipe may cross a deep valley, it will have the same nominal pressure, however each consumer will get a bit more or less because of the hydrostatic pressure (about 1 bar/10m height). So people at the bottom of a 100-foot (30 m) hill will get about 3 bars more than those at the top.

The effective pressure also varies because of the supply resistance even for the same static pressure. An urban consumer may have 5 metres of ½" lead pipe running from the iron main, so the kitchen tap flow will be fairly unrestricted, so high flow. A rural consumer may have a kilometre of rusted and limed ¾" iron pipe so their kitchen tap flow will be small.

For this reason the UK domestic water system has traditionally (prior to 1989) employed a "cistern feed" system, where the incoming supply is connected to the kitchen sink and also a header/storage tank in the attic. Water can dribble into this tank through a ½" lead pipe, plus ball valve, and then supply the house on 22 or 28 mm pipes. Gravity water has a small pressure (say ¼ bar in the bathroom) but needs wide pipes allow higher flows. This is fine for baths and toilets but is frequently inadequate for showers. People install shower booster pumps to increase the pressure. For this reason urban houses are increasingly using mains pressure boilers (combies) which take a long time to fill a bath but suit the high back pressure of a shower.

Comparing the Performance of Water and Sanitation Service Providers

Comparing the performance of water and sanitation service providers (utilities) is needed, because the sector offers limited scope for direct competition (natural monopoly). Firms operating in competitive markets are under constant pressure to out perform each other. Water utilities are often sheltered from this pressure, and it frequently shows: some utilities are on a sustained improvement track, but many others keep falling further behind best practice. Benchmarking the performance of utilities allows to simulate competition, establish realistic targets for improvement and create pressure to catch up with better utilities. Information on benchmarks for water and sanitation utilities is provided by the International Benchmarking Network for Water and Sanitation Utilities.

Institutional Responsibility and Governance

A great variety of institutions have responsibilities in water supply. A basic distinction is between institutions responsible for policy and regulation on the one hand; and institutions in charge of providing services on the other hand.

Policy and Regulation

Water supply policies and regulation are usually defined by one or several Ministries, in consultation with the legislative branch. In the United States the United States Environmental Protection Agencyý, whose administrator reports directly to the President, is responsible for water and sanitation policy and standard setting within the executive branch. In other countries responsibility for sector policy is entrusted to a Ministry of Environment (such as in Mexico and Colombia), to a Ministry of Health (such as in Panama, Honduras and Uruguay), a Ministry of Public Works (such as in Ecuador and Haiti), a Ministry of Economy (such as in German states) or a Ministry of Energy (such as in Iran). A few countries, such as Jordan and Bolivia, even have a Ministry of Water. Often several Ministries share responsibilities for water supply. In the European Union, important policy functions have been entrusted to the supranational level. Policy and regulatory functions include the setting of tariff rules and the approval of tariff increases; setting, monitoring and enforcing norms for quality of service and environmental protection; benchmarking the performance of service providers; and reforms in the structure of institutions responsible for service provision. The distinction between policy functions and regulatory functions is not always clear-cut. In some countries they are both entrusted to Ministries, but in others regulatory functions are entrusted to agencies that are separate from Ministries.

Regulatory Agencies

Dozens of countries around the world have established regulatory agencies for infrastructure services, including often water supply and sanitation, in order to better protect consumers and to improve efficiency. Regulatory agencies can be entrusted with a variety of responsibilities, including in particular the approval of tariff increases and the management of sector information systems, including benchmarking systems. Sometimes they also have a mandate to settle complaints by consumers that have not been dealt with satisfactorily by service providers. These specialised entities are expected to be more competent and objective in regulating service providers than departments of government Ministries. Regulatory agencies are supposed to be autonomous from the executive branch of government, but in many countries have often not been able to exercise a great degree of autonomy. In the United States regulatory agencies for utilities have existed for almost a century at the level of states, and in Canada at the level of provinces. In both countries they cover several infrastructure sectors. In many US states they are called Public Utility Commissions. For England and Wales, a regulatory agency for water (OFWAT) was created as part of the privatisation of the water industry in 1989. In many developing countries, water regulatory agencies were created during the 1990s in parallel with efforts at increasing private sector participation. Many countries do not have regulatory agencies for water. In these countries service providers are regulated directly by local government, or the national government. This is, for example, the case in the countries of continental Europe, in China and India.

Service Provision

Water supply service providers, which are often utilities, differ from each other in terms of their geographical coverage relative to administrative boundaries; their sectoral coverage; their ownership structure; and their governance arrangements.

Geographical Coverage

Many water utilities provide services in a single city, town or municipality. However, in many countries municipalities have associated in regional or inter-municipal or multi-jurisdictional utilities to benefit from economies of scale. In the United States these can take the form of special-purpose districts which may have independent taxing authority.

Figure: *The sole water supply of this section of Wilder, Tennessee, 1942.*

An example of a multi-jurisdictional water utility in the United States is WASA, a utility serving Washington, DC and various localities in the state of Maryland. Multi-jurisdictional utilities are also common in Germany, where they are known as "Zweckverbaende", in France and in Italy.

In some federal countries there are water service providers covering most or all cities and towns in an entire state, such as in all states of Brazil and some states in Mexico. In England and Wales water supply and sewerage is supplied almost entirely through ten regional companies. Some smaller countries, especially developed countries, have established service providers that cover the entire country or at least most of its cities and major towns. Such national service providers are especially prevalent in West Africa and Central America, but also exist, for example, in Tunisia, Jordan and Uruguay. In rural areas, where about half the world population lives, water services are often not provided by utilities, but by community-based organisations which usually cover one or sometimes several villages.

Sector Coverage

Some water utilities provide only water supply services, while sewerage is under the responsibility of a different entity. This is for example the case in Tunisia. However, in most cases water utilities also provide sewer and wastewater treatment services. In some cities or countries utilities also distribute electricity. In a few cases such multi-utilities also collect solid waste and provide local telephone services. An example of such an integrated utility can be found in the Colombian city of Medellín. Utilities that provide water, sanitation and electricity can be found in Frankfurt, Germany (Mainova), in Casablanca, Morocco and in Gabon in West Africa. Multi-utilities provide certain benefits such as common billing and the option to cross-subsidize water services with revenues from electricity sales, if permitted by law.

Ownership and Governance Arrangements

Water supply providers can be either public, private, mixed or cooperative. Most urban water supply services around the world are provided by public entities.

As Willem-Alexander, Prince of Orange (2002) stated, "The water crisis that is affecting so many people is mainly a crisis of governance—not of water scarcity." The introduction of cost-reflective tariffs together with cross-subsidisation between richer and poorer consumers is an essential governance reform in order to reduce the high levels of

Unaccounted or Water (UAW) and to provide the finance needed to extend the network to those poorest households who remain unconnected. Partnership arrangements between the public and private sector can play an important role in order to achieve this objective

Private Sector Participation

An estimated 10 percent of urban water supply is provided by private or mixed public-private companies, usually under concessions, leases or management contracts. Under these arrangements the public entity that is legally responsible for service provision delegates certain or all aspects of service provision to the private service provider for a period typically ranging from 4 to 30 years. The public entity continues to own the assets. These arrangements are common in France and in Spain. Only in few parts of the world water supply systems have been completely sold to the private sector (privatisation), such as in England and Wales as well as in Chile. The largest private water companies in the world are Suez and Veolia Environment from France; Aguas de Barcelona from Spain; and Thames Water from the UK, all of which are engaged internationally.

Public Water Service Provision

90% of urban water supply and sanitation services are currently in the public sector. They are owned by the state or local authorities, or also by collectives or cooperatives. They run without an aim for profit but are based on the ethos of providing a common good considered to be of public interest. In most middle and low-income countries, these publicly-owned and managed water providers can be inefficient as a result of political interference, leading to over-staffing and low labour productivity. Ironically, the main losers from this institutional arrangement are the urban poor in these countries. Because they are not connected to the network, they end up paying far more per litre of water than do more well-off households connected to the network who benefit from the implicit subsidies that they receive from loss-making utilities. We are still so far from achieving universal access to clean water and sanitation shows that public water authorities, in their current state, are not working well enough. Yet some are being very successful and are modelling the best forms of public management. As Ryutaro Hashimoto, former Japanese Prime Minister, notes: "Public water services currently provide more than 90 per cent of water supply in the world. Modest improvement in public water operators will have immense impact on global provision of services."

Governance Arrangements

Governance arrangements for both public and private utilities can take many forms. Governance arrangements define the relationship between the service provider, its owners, its customers and regulatory entities. They determine the financial autonomy of the service provider and thus its ability to maintain its assets, expand services, attract and retain qualified staff, and ultimately to provide high-quality services. Key aspects of governance arrangements are the extent to which the entity in charge of providing services is insulated from arbitrary political intervention; and whether there is an explicit mandate and political will to allow the service provider to recover all or at least most of its costs through tariffs and retain these revenues. If water supply is the responsibility of a department that is integrated in the administration of a city, town or municipality, there is a risk that tariff revenues are diverted for other purposes. In some cases, there is also a risk that staff are appointed mainly on political grounds rather than based on their professional credentials.

Tariffs

Almost all service providers in the world charge tariffs to recover part of their costs. According to estimates by the World Bank the average (mean) global water tariff is US$ 0.53 per cubic metre. In developed countries the average tariff is US$ 1.04, while it is only U$ 0.11 in the poorest developing countries. The lowest tariffs in developing countries are found in South Asia (mean of US$ 0.09/m3), while the highest are found in Latin America (US$ 0.41/m3). Data for 132 cities were assessed. The tariff is estimate for a consumption level of 15 cubic metres per month. Few utilities do recover all their costs. According to the same World Bank study only 30% of utilities globally, and only 50% of utilities in developed countries, generate sufficient revenue to cover operation, maintenance and partial capital costs.

According to another study undertaken in 2006 by NUS Consulting, the average water and sewerage tariff in 14 mainly OECD countries excluding VAT varied between US$ 0.66 per cubic metre in the United States and the equivalent of US$ 2.25 per cubic metre in Denmark. However, water consumption is much higher in the US than in Europe. Therefore, residential water bills may be very similar, even if the tariff per unit of consumption tends to be higher in Europe than in the US.

A typical family on the US East Coast paid between US$30 and US$70 per month for water and sewer services in 2005. In developing countries tariffs are usually much further from covering costs. Residential water bills for a typical consumption of 15 cubic metres per month vary between less than US$ 1 and US$ 12 per month.

Water and sanitation tariffs, which are almost always billed together, can take many different forms. Where metres are installed, tariffs are typically volumetric (per usage), sometimes combined with a small monthly fixed charge. In the absence of metres, flat or fixed rates — which are independent of actual consumption — are being charged. In developed countries, tariffs are usually the same for different categories of users and for different levels of consumption.

In developing countries, the situation is often characterised by cross-subsidies with the intent to make water more affordable for residential low-volume users that are assumed to be poor. For example, industrial and commercial users are often charged higher tariffs than public or residential users. Also, metred users are often charged higher tariffs for higher levels of consumption (increasing-block tariffs). However, cross-subsidies between residential users do not always reach their objective. Given the overall low level of water tariffs in developing countries even at higher levels of consumption, most consumption subsidies benefit the wealthier segments of society. Also, high industrial and commercial tariffs can provide an incentive for these users to supply water from other sources than the utility (own wells, water tankers) and thus actually erode the utility's revenue base.

Metring

Figure: *A typical residential water meter*

Metring of water supply is usually motivated by one or several of four objectives: First, it provides an incentive to conserve water which protects water resources (environmental objective). Second, it can postpone costly system expansion and saves energy and chemical costs (economic objective). Third, it allows a utility to better locate distribution losses (technical objective). Fourth, it allows to charge for water based on use, which is perceived by many as the fairest way to allocate the costs of water supply to users. Metring is considered good practice in water supply and is widespread in developed countries, except for the United Kingdom. In developing countries it is estimated that half of all urban water supply systems are metred and the tendency is increasing.

Water metres are read by one of several methods:

- the water customer writes down the metre reading and mails in a postcard with this info to the water department;
- the water customer writes down the metre reading and uses a phone dial-in system to transfer this info to the water department;
- the water customer logs in to the website of the water supply company, enters the address, metre ID and metre readings
- a metre reader comes to the premise and enters the metre reading into a handheld computer;
- the metre reading is echoed on a display unit mounted to the outside of the premise, where a metre reader records them;
- a small radio is hooked up to the metre to automatically transmit readings to corresponding receivers in handheld computers, utility vehicles or distributed collectors
- a small computer is hooked up to the metre that can either dial out or receive automated phone calls that give the reading to a central computer system.

Most cities are increasingly installing Automatic Metre Reading (AMR) systems to prevent fraud, to lower ever-increasing labour and liability costs and to improve customer service and satisfaction.

Costs and Financing

The cost of supplying water consists to a very large extent of fixed costs (capital costs and personnel costs) and only to a small extent of variable costs that depend on the amount of water consumed (mainly energy and chemicals). The full cost of supplying water in urban areas in developed countries is about US$1–2 per cubic metre depending on local costs and local water consumption levels. The cost of sanitation (sewerage and wastewater treatment) is another US$1–2 per cubic metre. These costs are somewhat lower in developing countries. Throughout the world, only part of these costs is usually billed to consumers, the remainder being financed through direct or indirect subsidies from local, regional or national governments. Besides subsidies water supply investments are financed through internally generated revenues as well as through debt. Debt financing can take the form of credits from commercial Banks, credits from international financial institutions such as the World Bank and regional development banks (in the case of developing countries), and bonds (in the case of some developed countries and some upper middle-income countries).

History

Throughout history people have devised systems to make getting and using water more convenient. Early Rome had indoor plumbing, meaning a system of aqueducts and pipes that terminated in homes and at public wells and fountains for people to use. London water supply infrastructure developed over many centuries from early mediaeval conduits, through major 19th century treatment works built in response to cholera threats, to modern large scale reservoirs.

Figure: *Chelsea Waterworks, 1752. Two Newcomen beam engines pumped Thames water from a canal to reservoirs at Green Park and Hyde Park.*

Water towers appeared around the late 19th century, as building height rose, and steam, electric and diesel-powered water pumps became available. As skyscrapers appeared, they needed rooftop water towers.

The technique of purification of drinking water by use of compressed liquefied chlorine gas was developed in 1910 by U.S. Army

Major (later Brig. Gen.) Carl Rogers Darnall(1867–1941), Professor of Chemistry at the Army Medical School. Shortly thereafter, Major (later Col.) William J. L. Lyster (1869–1947) of the Army Medical Department used a solution of calcium hypochlorite in a linen bag to treat water. For many decades, Lyster's method remained the standard for U.S. ground forces in the field and in camps, implemented in the form of the familiar Lyster Bag (also spelled Lister Bag). Darnall's work became the basis for present day systems of municipal water '*purification*'. Desalination appeared during the late 20th century, and is still limited to a few areas. During the beginning of the 21st Century, especially in areas of urban and suburban population centres, traditional centralised infrastructure have not been able to supply sufficient quantities of water to keep up with growing demand. Among several options that have been managed are the extensive use of desalination technology, this is especially prevalent in coastal areas and in "dry" countries like Australia. Decentralisation of water infrastructure has grown extensively as a viable solution including Rainwater harvesting and Stormwater harvesting where policies are eventually tending towards a more rational use and sourcing of water incorporation concepts such as "Fit for Purpose".

Construction Engineering

Construction engineering involves planning and execution of the designs from transportation, site development, hydraulic, environmental, structural and geotechnical engineers. As construction firms tend to have higher business risk than other types of civil engineering firms, many construction engineers tend to take on a role that is more business-like in nature: drafting and reviewing contracts, evaluating logistical operations, and closely monitoring prices of necessary supplies.

Earthquake Engineering

Earthquake engineering covers ability of various structures to withstand hazardous earthquake exposures at the sites of their particular location. Earthquake engineering is a sub discipline of the broader category of Structural engineering. The main objectives of earthquake engineering are:

- Understand interaction of structures with the shaky ground.

  Foresee the consequences of possible earthquakes.
- Design, construct and maintain structures to perform at earthquake exposure up to the expectations and in compliance with building codes.

Environmental Engineering

Environmental engineering deals with the treatment of chemical, biological, and/or thermal waste, the purification of water and air, and the remediation of contaminated sites, due to prior waste disposal or accidental contamination. Among the topics covered by environmental engineering are pollutant transport, water purification, waste water treatment, air pollution, solid waste treatment and hazardous waste management. Environmental engineers can be involved with pollution reduction, green engineering, and industrial ecology. Environmental engineering also deals with the gathering of information on the environmental consequences of proposed actions and the assessment of effects of proposed actions for the purpose of assisting society and policy makers in the decision making process.

Environmental engineering is the contemporary term for sanitary engineering, though sanitary engineering traditionally had not included much of the hazardous waste management and environmental remediation work covered by the term *environmental engineering*. Some other terms in use are public health engineering and environmental health engineering.

Geotechnical Engineering

Geotechnical engineering is an area of civil engineering concerned with the rock and soil that civil engineering systems are supported by. Knowledge from the fields of geology, material science and testing, mechanics, and hydraulics are applied by geotechnical engineers to safely and economically design foundations, retaining walls, and similar structures. Environmental concerns in relation to groundwater and waste disposal have spawned a new area of study called geoenvironmental engineering where biology and chemistry are important.

Some of the unique difficulties of geotechnical engineering are the result of the variability and properties of soil. Boundary conditions are often well defined in other branches of civil engineering, but with soil, clearly defining these conditions can be impossible. The material properties and behaviour of soil are also difficult to predict due to the variability of soil and limited investigation. This contrasts with the relatively well defined material properties of steel and concrete used in other areas of civil engineering. Soil mechanics, which describes the behaviour of soil, is also complicated because soils exhibit nonlinear (stress-dependent) strength, stiffness, and dilatancy (volume change associated with application of shear stress).

Water Resources Engineering

Water resources engineering is concerned with the collection and management of water (as a natural resource). As a discipline it therefore combines hydrology, environmental science, meteorology, geology, conservation, and resource management. This area of civil engineering relates to the prediction and management of both the quality and the quantity of water in both underground (aquifers) and above ground (lakes, rivers, and streams) resources. Water resource engineers analyse and model very small to very large areas of the earth to predict the amount and content of water as it flows into, through, or out of a facility. Although the actual design of the facility may be left to other engineers. Hydraulic engineering is concerned with the flow and conveyance of fluids, principally water. This area of civil engineering is intimately related to the design of pipelines, water supply network, drainage facilities (including bridges, dams, channels, culverts, levees, storm sewers), and canals. Hydraulic engineers design these facilities using the concepts of fluid pressure, fluid statics, fluid dynamics, and hydraulics, among others.

Figure: *Hoover dam*

Materials Engineering

Another aspect of Civil engineering is materials science. Material engineering deals with ceramics such as concrete, mix asphalt concrete, metals Focus around increased strength, metals such as aluminium and steel, and polymers such as polymethylmethacrylate (PMMA) and carbon fibres.

Materials engineering also consists of protection and prevention like paints and finishes. Alloying is another aspect of material engineering, combining two different types of metals to produce a stronger metal.

Surveying

Surveying is the process by which a surveyor measures certain dimensions that generally occur on the surface of the Earth. Surveying equipment, such as levels and theodolites, are used for accurate measurement of angular deviation, horizontal, vertical and slope distances. With computerisation, electronic distance measurement (EDM), total stations, GPS surveying and laser scanning have supplemented (and to a large extent supplanted) the traditional optical instruments. This information is crucial to convert the data into a graphical representation of the Earth's surface. This information is then used by civil engineers, contractors and even realtors to design from, build on, and trade, respectively. Elements of a building or structure must be correctly sized and positioned in relation to each other and to site boundaries and adjacent structures. Although surveying is a distinct profession with separate qualifications and licensing arrangements, civil engineers are trained in the basics of surveying and mapping, as well as geographic information systems. Surveyors may also lay out the routes of railways, tramway tracks, highways, roads, pipelines and streets as well as position other infrastructures, such as harbors, before construction.

Land Surveying

In the United States, Canada, the United Kingdom and most Commonwealth countries land surveying is considered to be a distinct profession. Land surveyors are not considered to be engineers, and have their own professional associations and licencing requirements. The services of a licenced land surveyor are generally required for boundary surveys (to establish the boundaries of a parcel using its legal description) and subdivision plans (a plot or map based on a survey of a parcel of land, with boundary lines drawn inside the larger parcel to indicate the creation of new boundary lines and roads), both of which are generally referred to as cadastral surveying.

Construction Surveying

Construction surveying is generally performed by specialised technicians. Unlike land surveyors, the resulting plan does not have legal status. Construction surveyors perform the following tasks:

- Survey existing conditions of the future work site, including topography, existing buildings and infrastructure, and even including underground infrastructure whenever possible;
- Construction surveying (otherwise "lay-out" or "setting-out"): to stake out reference points and markers that will guide the

construction of new structures such as roads or buildings for subsequent construction;

- Verify the location of structures during construction;
- As-Built surveying: a survey conducted at the end of the construction project to verify that the work authorised was completed to the specifications set on plans.

Transportation Engineering

Transportation engineering is concerned with moving people and goods efficiently, safely, and in a manner conducive to a vibrant community. This involves specifying, designing, constructing, and maintaining transportation infrastructure which includes streets, canals, highways, rail systems, airports, ports, and mass transit. It includes areas such as transportation design, transportation planning, traffic engineering, some aspects of urban engineering, queueing theory, pavement engineering, Intelligent Transportation System (ITS), and infrastructure management.

Municipal or Urban Engineering

Municipal engineering is concerned with municipal infrastructure. This involves specifying, designing, constructing, and maintaining streets, sidewalks, water supply networks, sewers, street lighting, municipal solid waste management and disposal, storage depots for various bulk materials used for maintenance and public works (salt, sand, etc.), public parks and bicycle paths.

In the case of underground utility networks, it may also include the civil portion (conduits and access chambers) of the local distribution networks of electrical and telecommunications services. It can also include the optimising of waste collection and bus service networks. Some of these disciplines overlap with other civil engineering specialities, however municipal engineering focuses on the coordination of these infrastructure networks and services, as they are often built simultaneously, and managed by the same municipal authority.

Water Supply System

A water supply system or water supply network is a system of engineered hydrologic and hydraulic components which provide water supply. A water supply system typically includes:

1. A drainage basin;
2. A raw (untreated) water collection point (above or below ground) where the water accumulates, such as a lake, a river, or

groundwater from an underground aquifer. Untreated drinking water (usually water being transferred to the water purification facilities) may be transferred using uncovered ground-level aqueducts, covered tunnels or underground water pipes.

3. Water purification facilities. Treated water is transferred using water pipes (usually underground).
4. Water storage facilities such as reservoirs, water tanks, or watertowers. Smaller water systems may store the water in cisterns or pressure vessels. (Tall buildings may also need to store water locally in pressure vessels in order for the water to reach the upper floors.)
5. Additional water pressurising components such as pumping stations may need to be situated at the outlet of underground or above ground reservoirs or cisterns (if gravity flow is impractical)
6. A pipe network for distribution of water to the consumers (which may be private houses or industrial, commercial or institution establishments) and other usage points (such as fire hydrants)
7. Connections to the sewers (underground pipes, or aboveground ditches in some developing countries) are generally found downstream of the water consumers, but the sewer system is considered to be a separate system, rather than part of the water supply system.

Water Abstraction and Raw Water Transfer

Raw water (untreated) is collected from a surface water source (such as an intake on a lake or a river) or from a groundwater source (such as a water well drawing from an underground aquifer) within the watershed that provides the water resource.

Shallow dams and reservoirs are susceptible to outbreaks of toxic algae, especially if the water is warmed by a hot sun. The bacteria grow from stormwater runoff carrying fertilizer into the river where it acts as a nutrient for the algae. Such outbreaks render the water unfit for human consumption. The raw water is transferred to the water purification facilities using uncovered aqueducts, covered tunnels or underground water pipes.

Water Treatment

Virtually all large systems must treat the water; a fact that is tightly regulated by global, state and federal agencies, such as the

World Health Organisation (WHO) or the United States Environmental Protection Agency (EPA). Water treatment must occur before the product reaches the consumer and afterwards (when it is discharged again). Water purification usually occurs close to the final delivery points to reduce pumping costs and the chances of the water becoming contaminated after treatment.

Traditional surface water treatment plants generally consists of three steps: clarification, filtration and disinfection. Clarification refers to the separation of particles (dirt, organic matter, etc.) from the water stream. Chemical addition (i.e. alum, ferric chloride) destabilises the particle charges and prepares them for clarification either by settling or floating out of the water stream. Sand, anthracite or activated carbon filters refine the water stream, removing smaller particulate matter. While other methods of disinfection exist, the preferred method is via chlorine addition. Chlorine effectively kills bacteria and most viruses and maintains a residual to protect the water supply through the supply network.

Water Distribution Network

The product, delivered to the point of consumption, is called fresh water if it receives little or no treatment, or drinking water if the treatment achieves the water quality standards required for human consumption.

Figure: *Most (treated) water distribution happens through underground pipes*

Once treated, chlorine is added to the water and it is distributed by the local supply network. Today, water supply systems are typically constructed of plastic, ferrous, or concrete circular pipe. However, other "pipe" shapes and material may be used, such as square or rectangular concrete boxes, arched brick pipe, or wood. Near the end point, the network of pipes through which the water is delivered is often referred to as the *water mains*.

The energy that the system needs to deliver the water is called pressure. That energy is transferred to the water, therefore becoming water pressure, in a number of ways: by a pump, by gravity feed from a water source (such as a water tower) at a higher elevation, or by compressed air. The water is often transferred from a water reserve such as a large communal reservoir before being transported to a more pressurised reserve such as a watertower.

In small domestic systems, the water may be pressurised by a pressure vessel or even by an underground cistern (the latter however does need additional pressurising). This eliminates the need of a water-tower or any other heightened water reserve to supply the water pressure.

These systems are usually owned and maintained by local governments, such as cities, or other public entities, but are occasionally operated by a commercial enterprise. Water supply networks are part of the master planning of communities, counties, and municipalities. Their planning and design requires the expertise of city planners and civil engineers, who must consider many factors, such as location, current demand, future growth, leakage, pressure, pipe size, pressure loss, fire fighting flows, etc. — using pipe network analysis and other tools. Construction comparable sewage systems, was one of the great engineering advances that made urbanisation possible. Improvement in the quality of the water has been one of the great advances in public health. As water passes through the distribution system, the water quality can degrade by chemical reactions and biological processes. Corrosion of metal pipe materials in the distribution system can cause the release of metals into the water with undesirable aesthetic and health effects. Release of iron from unlined iron pipes can result in customer reports of "red water" at the tap. Release of copper from copper pipes can result in customer reports of "blue water" and/or a metallic taste. Release of lead can occur from the solder used to join copper pipe together or from brass fixtures. Copper and lead levels at the consumer's tap are regulated to protect consumer health.

Utilities will often adjust the chemistry of the water before distribution to minimise its corrosiveness. The simplest adjustment involves control of pH and alkalinity to produce a water that tends to passivate corrosion by depositing a layer of calcium carbonate. Corrosion inhibitors are often added to reduce release of metals into the water. Common corrosion inhibitors added to the water are phosphates and silicates. Maintenance of a biologically safe drinking water is another goal in water distribution. Typically, a chlorine based disinfectant, such as sodium hypochlorite or monochloramine is added to the water as it leaves the treatment plant. Booster stations can be placed within the distribution system to ensure that all areas of the distribution system have adequate sustained levels of disinfection.

Topologies of Water Distribution Networks

Like electric power lines, roads, and microwave radio networks, water systems may have a loop or branch network topology, or a combination of both. The piping networks are circular or rectangular. If any one section of water distribution main fails or needs repair, that section can be isolated without disrupting all users on the network. Most systems are divided into zones. Factors determining the extent or size of a zone can include hydraulics, telemetry systems, history, and population density. Sometimes systems are designed for a specific area then are modified to accommodate development. Terrain affects hydraulics and some forms of telemetry. While each zone may operate as a stand-alone system, there is usually some arrangement to interconnect zones in order to manage equipment failures or system failures.

Water Network Maintenance

Water supply networks usually represent the majority of assets of a water utility. Systematic documentation of maintenance works using a Computerized Maintenance Management System is a key to a successful operation of a water utility.

Structural Engineering

Structural engineering is a field of engineering dealing with the analysis and design of structures that support or resist loads. Structural engineering is usually considered a speciality within civil engineering, but it can also be studied in its own right. Structural engineers are most commonly involved in the design of buildings and large nonbuilding structures but they can also be involved in the design of machinery, medical equipment, vehicles or any item where structural integrity affects the item's function or safety. Structural engineers must ensure their designs satisfy given design criteria, predicated on safety (e.g. structures must not collapse without due warning) or serviceability and performance (e.g. building sway must not cause discomfort to the occupants). Buildings are made to endure massive loads as well as changing climate and natural disasters.

Structural engineering theory is based upon physical laws and empirical knowledge of the structural performance of different landscapes and materials. Structural engineering design utilises a relatively small number of basic structural elements to build up structural systems that can be very complex. Structural engineers are responsible for making creative and efficient use of funds, structural elements and materials to achieve these goals.

A Structural Engineer

Structural engineers are responsible for engineering design and analysis. Entry-level structural engineers may design the individual structural elements of a structure, for example the beams, columns, and floors of a building. More experienced engineers would be responsible for the structural design and integrity of an entire system, such as a building. Structural engineers often specialise in particular fields, such as bridge engineering, building engineering, pipeline engineering, industrial structures, or special mechanical structures such as vehicles or aircraft.

Structural engineering has existed since humans first started to construct their own structures. It became a more defined and formalised profession with the emergence of the architecture profession as distinct from the engineering profession during the industrial revolution in the late 19th century. Until then, the architect and the structural engineer were usually one and the same - the master builder. Only with the development of specialised knowledge of structural theories that emerged during the 19th and early 20th centuries did the professional structural engineer come into existence.

The role of a structural engineer today involves a significant understanding of both static and dynamic loading, and the structures that are available to resist them. The complexity of modern structures often requires a great deal of creativity from the engineer in order to ensure the structures support and resist the loads they are subjected to. A structural engineer will typically have a four or five year undergraduate degree, followed by a minimum of three years of professional practice before being considered fully qualified.

Structural engineers are licensed or accredited by different learned societies and regulatory bodies around the world (for example, the Institution of Structural Engineers in the UK). Depending on the degree course they have studied and/or the jurisdiction they are seeking licensure in, they may be accredited (or licensed) as just structural engineers, or as civil engineers, or as both civil and structural engineers.

History of Structural Engineering

Structural engineering dates back to 2700 BC when the step pyramid for Pharaoh Djoser was built by Imhotep, the first engineer in history known by name. Pyramids were the most common major structures built by ancient civilizations because the structural form of a pyramid is inherently stable and can be almost infinitely scaled (as opposed to most other structural forms, which cannot be linearly increased in size in proportion to increased loads).

Throughout ancient and medieval history most architectural design and construction was carried out by artisans, such as stone masons and carpenters, rising to the role of master builder. No theory of structures existed, and understanding of how structures stood up was extremely limited, and based almost entirely on empirical evidence of 'what had worked before'. Knowledge was retained by guilds and seldom supplanted by advances. Structures were repetitive, and increases in scale were incremental.

No record exists of the first calculations of the strength of structural members or the behaviour of structural material, but the profession of structural engineer only really took shape with the industrial revolution and the re-invention of concrete. The physical sciences underlying structural engineering began to be understood in the Renaissance and have been developing ever since.

Structural Failure

The history of structural engineering contains many collapses and failures. Sometimes this is due to obvious negligence, as in the case of the Pétionville school collapse, in which Rev. Fortin Augustin said that *"he constructed the building all by himself, saying he didn't need an engineer as he had good knowledge of construction"* following a partial collapse of the three-story schoolhouse that sent neighbours fleeing. The final collapse killed at least 362 people, mostly children.

In other cases structural failures require careful study, and the results of these inquiries have resulted in improved practices and greater understanding of the science of structural engineering. Some such studies are the result of Forensic engineering investigations where the original engineer seems to have done everything in accordance with the state of the profession and acceptable practice yet a failure still eventuated. A famous case of structural knowledge and practice being advanced in this manner can be found in a series of failures involving Box girders which collapsed in Australia during the 1970s.

Specialisations

Building Structures: Structural building engineering includes all structural engineering related to the design of buildings. It is the branch of structural engineering that is close to architecture.

Structural building engineering is primarily driven by the creative manipulation of materials and forms and the underlying mathematical and scientific ideas to achieve an end which fulfills its functional requirements and is structurally safe when subjected to all the loads it could reasonably be expected to experience. This is subtly different from architectural design, which is driven by the creative manipulation of materials and forms, mass, space, volume, texture and light to achieve an end which is aesthetic, functional and often artistic.

The architect is usually the lead designer on buildings, with a structural engineer employed as a sub-consultant. The degree to which each discipline actually leads the design depends heavily on the type of structure. Many structures are structurally simple and led by architecture, such as multi-storey office buildings and housing, while other structures, such as tensile structures, shells and gridshells are heavily dependent on their form for their strength, and the engineer may have a more significant influence on the form, and hence much of the aesthetic, than the architect.

The structural design for a building must ensure that the building is able to stand up safely, able to function without excessive deflections or movements which may cause fatigue of structural elements, cracking or failure of fixtures, fittings or partitions, or discomfort for occupants. It must account for movements and forces due to temperature, creep, cracking and imposed loads. It must also ensure that the design is practically buildable within acceptable manufacturing tolerances of the materials. It must allow the architecture to work, and the building services to fit within the building and function (air conditioning, ventilation, smoke extract, electrics, lighting etc.). The structural design of a modern building can be extremely complex, and often requires a large team to complete.

Structural engineering specialities for buildings include:

- Earthquake engineering
- Façade engineering
- Fire engineering
- Roof engineering

- Tower engineering
- Wind engineering.

Earthquake Engineering Structures

Earthquake engineering structures are those engineered to withstand various types of hazardous earthquake exposures at the sites of their particular location.

Earthquake engineering is treating its subject structures like defensive fortifications in military engineering but for the warfare on earthquakes. Both earthquake and military general design principles are similar: be ready to slow down or mitigate the advance of a possible attacker.

The main objectives of earthquake engineering are:

- Understand interaction of structures with the shaky ground.
- Foresee the consequences of possible earthquakes.
- Design and construct the structures to perform while being exposed to an earthquake.

Earthquake engineering or earthquake-proof structure does not, necessarily, mean *extremely strong* and *expensive* one like El Castillo pyramid at Chichen Itza. In fact, many structures considered *strong* may in fact be actually *stiff,* which may result in poor seismic performance.

Now, the most *powerful* and *budgetary* tool of the earthquake engineering is base isolation which pertains to the passive structural vibration control technologies.

Civil Engineering Structures

Civil structural engineering includes all structural engineering related to the built environment. It includes:

- Bridges
- Dams
- Earthworks
- Foundations
- Offshore structures
- Pipelines
- Power stations
- Railways
- Retaining structures and walls

- Roads
- Tunnels
- Waterways
- Water and wastewater infrastructure.

The structural engineer is the lead designer on these structures, and often the sole designer. In the design of structures such as these, structural safety is of paramount importance (in the UK, designs for dams, nuclear power stations and bridges must be signed off by a chartered engineer). Civil engineering structures are often subjected to very extreme forces, such as large variations in temperature, dynamic loads such as waves or traffic, or high pressures from water or compressed gases. They are also often constructed in corrosive environments, such as at sea, in industrial facilities or below ground.

Mechanical Structures

Principals of structural engineering are applied to variety of mechanical (moveable) structures. The design of static structures assumes they always have the same geometry (in fact, so-called static structures can move significantly, and structural engineering design must take this into account where necessary), but the design of moveable or moving structures must account for fatigue, variation in the method in which load is resisted and significant deflections of structures.

The forces which parts of a machine are subjected to can vary significantly, and can do so at a great rate. The forces which a boat or aircraft are subjected to vary enormously and will do so thousands of times over the structure's lifetime. The structural design must ensure that such structures are able to endure such loading for their entire design life without failing.

These works can require mechanical structural engineering:

- Airframes and fuselages
- Boilers and pressure vessels
- Coachworks and carriages
- Cranes
- Elevators
- Escalators
- Marine vessels and hulls.

Structural Elements

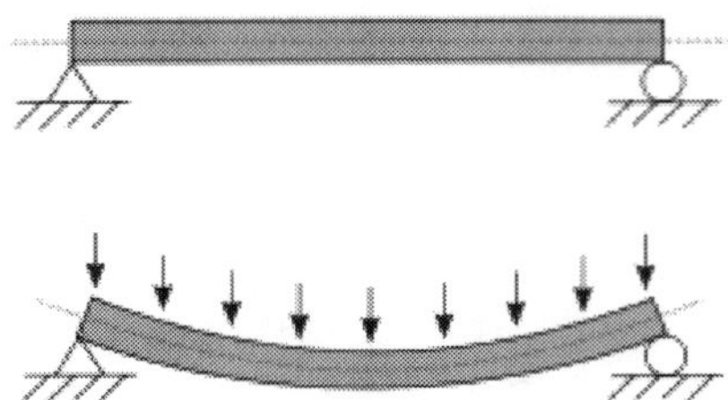

Figure: A statically determinate simply supported beam, bending under an evenly distributed load.

Any structure is essentially made up of only a small number of different types of elements:

- Columns
- Beams
- Plates
- Arches
- Shells
- Catenaries.

Columns

Columns are elements that carry only axial force - either tension or compression - or both axial force and bending (which is technically called a beam-column but practically, just a column). The design of a column must check the axial capacity of the element, and the buckling capacity.

The buckling capacity is the capacity of the element to withstand the propensity to buckle. Its capacity depends upon its geometry, material, and the effective length of the column, which depends upon the restraint conditions at the top and bottom of the column. The effective length is $K * l$ where l is the real length of the column.

The capacity of a column to carry axial load depends on the degree of bending it is subjected to, and vice versa. This is represented on an interaction chart and is a complex non-linear relationship.

Beams

A beam may be defined as an element in which one dimemsion is much greater than the other two and the applied loads are usually normal to the main axis of the element. Beams and columns are called line elements and are often represented by simple lines in structural modelling.

- cantilevered (supported at one end only with a fixed connection)
- simply supported (supported vertically at each end; horizontally on only one to withstand friction, and able to rotate at the supports)
- continuous (supported by three or more supports)
- a combination of the above (ex. supported at one end and in the middle).

Beams are elements which carry pure bending only. Bending causes one part of the section of a beam (divided along its length) to go into compression and the other part into tension. The compression part must be designed to resist buckling and crushing, while the tension part must be able to adequately resist the tension.

Struts and Ties

A truss is a structure comprising two types of structural elements; compression members and tension members (i.e. struts and ties). Most trusses use gusset plates to connect intersecting elements. Gusset plates are relatively flexible and minimise bending moments at the connections, thus allowing the truss members to carry primarily tension or compression. Trusses are usually utilised in span large distances, where it would be uneconomical to use solid beams.

Plates

Plates carry bending in two directions. A concrete flat slab is an example of a plate. Plates are understood by using continuum mechanics, but due to the complexity involved they are most often designed using a codified empirical approach, or computer analysis.

They can also be designed with yield line theory, where an assumed collapse mechanism is analysed to give an upper bound on the collapse load. This is rarely used in practice.

Shells

Shells derive their strength from their form, and carry forces in compression in two directions. A dome is an example of a shell. They can be designed by making a hanging-chain model, which will act as a catenary in pure tension, and inverting the form to achieve pure compression.

Arches

Arches carry forces in compression in one direction only, which is why it is appropriate to build arches out of masonry. They are designed by ensuring that the line of thrust of the force remains within the depth of the arch. Its mainly used to increased the bountifulness of any structure.

Catenaries

Catenaries derive their strength from their form, and carry transverse forces in pure tension by deflecting (just as a tightrope will sag when someone walks on it). They are almost always cable or fabric structures. A fabric structure acts as a catenary in two directions.

Structural Engineering Theory

Structural engineering depends upon a detailed knowledge of applied mechanics, materials science and applied mathematics to understand and predict how structures support and resist self-weight and imposed loads. To apply the knowledge successfully a structural engineer generally requires detailed knowledge of relevant empirical and theoretical design codes, the techniques of structural analysis, as well as some knowledge of the corrosion resistance of the materials and structures, especially when those structures are exposed to the external environment. Since the 1990s, specialist software has become available to aid in the design of structures, with the functionality to assist in the drawing, analysing and designing of structures with maximum precision; examples include AutoCAD, StaadPro, ETABS, Prokon etc.

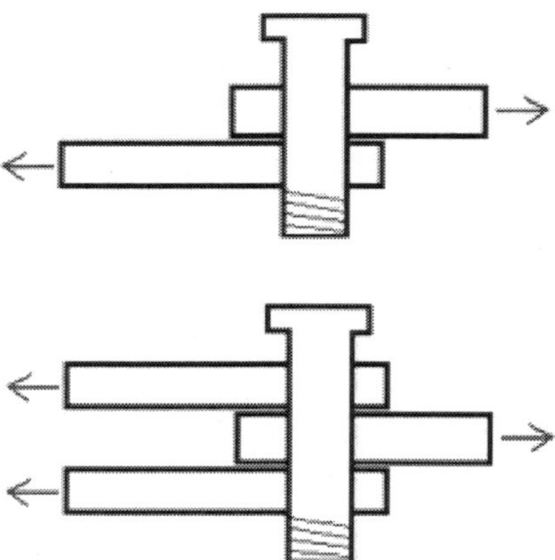

Figure: *Figure of a bolt in shear stress. Top figure illustrates single shear, bottom figure illustrates double shear.*

Such software may also take into consideration environmental loads, such as from earthquakes and winds.

7

Hydropower

Hydropower, hydraulic power, hydrokinetic power or water power is power that is derived from the force or energy of falling water, which may be harnessed for useful purposes. Since ancient times, hydropower has been used for irrigation and the operation of various mechanical devices, such as watermills, sawmills, textile mills, dock cranes, and domestic lifts. Since the early 20th century, the term is used almost exclusively in conjunction with the modern development of hydro-electric power, the energy of which could be transmitted considerable distance between where it was created to where it was consumed. Another previous method used to transmit energy had employed a trompe, which produces compressed air from falling water, that could then be piped to power other machinery at a distance from the energy source.

Water's power is manifested in hydrology, by the forces of water on the riverbed and banks of a river. When a river is in flood, it is at its most powerful, and moves the greatest amount of sediment. This higher force results in the removal of sediment and other material from the riverbed and banks of the river, locally causing erosion, transport and, with lower flow, sedimentation downstream. Early uses of waterpower date back to Mesopotamia and ancient Egypt, where irrigation has been used since the 6th millennium BC and water clocks had been used since the early 2nd millennium BC. Other early examples of water power include the Qanat system in ancient Persia and the Turpan water system in ancient China.

Waterwheels, Turbines, and Mills

Hydropower has been used for hundreds of years. In India, water wheels and watermills were built; in Imperial Rome, water powered mills produced flour from grain, and were also used for sawing timber and stone; in China, watermills were widely used since the Han Dynasty.

The power of a wave of water released from a tank was used for extraction of metal ores in a method known as hushing. The method was first used at the Dolaucothi gold mine in Wales from 75 AD onwards, but had been developed in Spain at such mines as Las Medulas. Hushing was also widely used in Britain in the Medieval and later periods to extract lead and tin ores. It later evolved into hydraulic mining when used during the California gold rush. In China and the rest of the Far East, hydraulically operated "pot wheel" pumps raised water into irrigation canals. At the beginning of the Industrial revolution in Britain, water was the main source of power for new inventions such as Richard Arkwright's water frame.

Although the use of water power gave way to steam power in many of the larger mills and factories, it was still used during the 18th and 19th centuries for many smaller operations, such as driving the bellows in small blast furnaces (e.g. the Dyfi Furnace) and gristmills, such as those built at Saint Anthony Falls, which uses the 50-foot (15 m) drop in the Mississippi River.

In the 1830s, at the early peak in U.S. canal-building, hydropower provided the energy to transport barge traffic up and down steep hills using inclined plane railroads. About the same time and later, as railroads overtook canals for transportation, canal systems were modified and developed into hydropower systems; the history of Lowell, Massachusetts is a classic example of commercial development and industrialisation, built upon the availability of water power.

During this time also, technological advances had moved the water wheel to a turbine, and in 1848 James B. Francis, while working as head engineer of Lowell's Locks and Canals company, improved on these designs to create a turbine with 90% efficiency. Applying scientific principles and testing methods he produced a very efficient turbine design, but more importantly, his mathematical and graphical calculation methods improved turbine design and engineering.

His analytical methods allowed confident design of high efficiency turbines to exactly match a site's specific flow conditions. The Francis reaction turbine is still in wide usage today, with few changes in either design or methods. In the 1870s and deriving from uses in the California mining industry, Lester Allan Pelton developed the high efficiency Pelton wheel impulse turbine, which utilised hydropower with very different conditions of flow and pressure.

Hydraulic Power-pipe Networks

Hydraulic power networks also developed, using pipes to carrying pressurised water and transmit mechanical power from the source to end users elsewhere locally; the power source was normally a head of water, which could also be assisted by a pump.

These were extensive in Victorian cities in the United Kingdom. A hydraulic power network was also developed in Geneva, Switzerland. The world famous Jet d'Eau was originally designed as the over-pressure relief valve for the network.

Compressed Air Hydro

Where there is a plentiful head of water it can be made to generate compressed air directly without moving parts, if the two phases are separated. In these designs, a falling column of water is purposely mixed with air bubbles generated through turbulence at the high level intake.

This is allowed to fall down a shaft into a subterranean, high-roofed chamber where the now-compressed air separates from the water and becomes trapped.

The height of falling water column maintains compression of the air in the top of the chamber, while an outlet, submerged below the water level in the chamber allows water to flow back to the surface at a slightly lower level than the intake. A separate outlet in the roof of the chamber supplies the compressed air to the surface. A facility on this principal was built on the Montreal River at Ragged Shutes near Cobalt, Ontario in 1910 and supplied 5,000 horsepower to nearby mines.

Modern Usage

There are several mature forms of water power currently in wide usage, and other forms which are still in phases of development. Most hydropower is used primarily to generate electricity, but some are purely mechanical. Broad categories include:

Riverine Hydropower

- Conventional hydroelectric, referring to hydroelectric dams.
- Run-of-the-river hydroelectricity, which captures the kinetic energy in rivers or streams, without the use of dams.

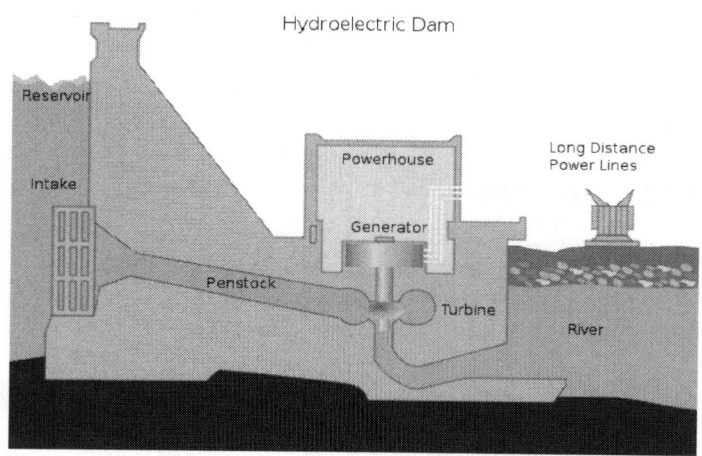

Figure: *A conventional dammed-hydro facility (hydroelectric dam) is the most common type of hydroelectric power generation.*

- Pumped-storage hydroelectricity, to pump up water, and use its head to generate in times of demand.

Marine Energy

- Marine current power, which captures the kinetic energy from marine currents.
- Osmotic power, which channels river water into a container separated from sea water by a semi-permeable membrane.
- Ocean thermal energy, which exploits the temperature difference between deep and shallow waters.
- Tidal power, which captures energy from the tides in horizontal direction. Also a popular form of hydroelectric power generation.
  - * Tidal stream power, usage of stream generators, somewhat similar to that of a wind turbine.
  - * Tidal barrage power, usage of a tidal dam.
  - * Dynamic tidal power, utilising large areas to generate head.
- Wave power, the use ocean surface waves to generate power.

Calculating the Amount of Available Power

A hydropower resource can be measured according to the amount of available power, or energy per unit time. In large reservoirs, the available power is generally only a function of the hydraulic head and rate of fluid flow. In a reservoir, the head is the height of water in the reservoir relative to its height after discharge. Each unit of water can do an amount of work equal to its weight times the head. The amount of energy, *E*, released when an object of mass *m* drops a height *h* in a gravitational field of strength *g* is given by

$$E = mgh$$

The energy available to hydroelectric dams is the energy that can be liberated by lowering water in a controlled way. In these situations, the power is related to the mass flow rate.

$$\frac{E}{t} = \frac{m}{t} gh$$

Substituting *P* for $^{E}D_{t}$ and expressing $^{m}D_{t}$ in terms of the volume of liquid moved per unit time (the rate of fluid flow, φ) and the density of water, we arrive at the usual form of this expression:

$$P = \rho\phi gh$$

or

A simple formula for approximating electric power production at a hydroelectric plant is:

$$P = hrgk$$

where P is Power in kilowatts, h is height in metres, r is flow rate in cubic metres per second, g is acceleration due to gravity of 9.8 m/s2, and k is a coefficient of efficiency ranging from 0 to 1. Efficiency is often higher with larger and more modern turbines. Some hydropower systems such as water wheels can draw power from the flow of a body of water without necessarily changing its height. In this case, the available power is the kinetic energy of the flowing water.

$$P = \frac{1}{2}\rho\phi v^2$$

where *v* is the speed of the water, or with

$$\phi = Av$$

where *A* is the area through which the water passes, also

$$P = \frac{1}{2}\rho A v^3$$

Over-shot water wheels can efficiently capture both types of energy.

Deep Water Source Cooling

Deep water source cooling (DWCS) or deep water air cooling is a form of air cooling for process and comfort space cooling which uses a renewable, large body of naturally cold water as a heat sink. It uses water at a constant 277 to 283 kelvins (4 to 10 degrees Celsius) or less which it withdraws from deep areas within lakes, oceans, aquifers and rivers and is pumped through the primary side of a heat exchanger. On the secondary side of the heat exchanger, cooled water is produced.

Advantages

Deep water source cooling has several advantages:

- It is very energy efficient, requiring only 1/10 of the average energy required by conventional cooler systems.
- It does not use any ozone depleting refrigerant. This is still an important feature, despite the fact that smaller/less expensive ecologic cooling devices also exist.

Disadvantages

Several disadvantages are also present:

- It requires the presence of a (fairly) large and deep water quantity in the surroundings. To obtain water in the 3 to 6 °C (37 to 43 °F) range, a depth of 66 metres (217 ft) is required.
- The set-up of a system is expensive and labour-intensive. The system also requires a great amount of source material for its construction and placement. For example, the pipes alone can range several hundreds of metres or feet when placed end to end.

Ocean Thermal Energy Conversion

Ocean thermal energy conversion (*OTEC*) uses the difference between cooler deep and warmer shallow or surface ocean waters to run a heat engine and produce useful work, usually in the form of electricity.

A heat engine gives greater efficiency and power when run with a large temperature difference. In the oceans the temperature difference between surface and deep water is greatest in the tropics, although still a modest 20°C to 25°C. It is therefore in the tropics that OTEC offers the greatest possibilities. OTEC has the potential to offer global amounts of energy that are 10 to 100 times greater than other ocean energy options such as wave power. OTEC plants can operate continuously providing a base load supply for an electrical power generation system.

The main technical challenge of OTEC is to generate significant amounts of power efficiently from small temperature differences. It is still considered an emerging technology. Early OTEC systems were of 1 to 3% thermal efficiency, well below the theoretical maximum for this temperature difference of between 6 and 7%. Current designs are expected to be closer to the maximum. The first operational system was built in Cuba in 1930 and generated 22 kW. Modern designs allow performance approaching the theoretical maximum Carnot efficiency and the largest built in 1999 by the USA generated 250 kW.

The most commonly used heat cycle for OTEC is the Rankine cycle using a low-pressure turbine. Systems may be either closed-cycle or open-cycle. Closed-cycle engines use working fluids that are typically thought of as refrigerants such as ammonia or R-134a. Open-cycle engines use vapour from the seawater itself as the working fluid.

OTEC can also supply quantities of cold water as a by-product. This can be used for air conditioning and refrigeration and the fertile deep ocean water can feed biological technologies. Another by-product is fresh water distilled from the sea.

Attempts to develop and refine OTEC technology started in the 1880s. In 1881, Jacques Arsene d'Arsonval, a French physicist, proposed tapping the thermal energy of the ocean. D'Arsonval's student, Georges Claude, built the first OTEC plant, in Matanzas, Cuba in 1930. The system generated 22 kW of electricity with a low-pressure turbine.

In 1931, Nikola Tesla released "Our Future Motive Power", which described such a system. Tesla ultimately concluded that the scale of engineering required made it impractical for large scale development.

In 1935, Claude constructed a plant aboard a 10,000-ton cargo vessel moored off the coast of Brazil. Weather and waves destroyed it before it could generate net power. (Net power is the amount of power generated after subtracting power needed to run the system.)

In 1956, French scientists designed a 3 MW plant for Abidjan, Ivory Coast. The plant was never completed, because new finds of large amounts of cheap oil made it uneconomical.

In 1962, J. Hilbert Anderson and James H. Anderson, Jr. focused on increasing component efficiency. They patented their new "closed cycle" design in 1967.

Japan is a major contributor to the development of the technology. Beginning in 1970 the Tokyo Electric Power Company successfully built and deployed a 100 kW closed-cycle OTEC plant on the island of Nauru. The plant became operational 1981-10-14, producing about 120 kW of electricity; 90 kW was used to power the plant and the remaining electricity was used to power a school and other places. This set a world record for power output from an OTEC system where the power was sent to a real power grid.

The United States became involved in 1974, establishing the Natural Energy Laboratory of Hawaii Authority at Keahole Point on the Kona coast of Hawai»i. Hawaii is the best U.S. OTEC location, due to its warm surface water, access to very deep, very cold water, and Hawaii's high electricity costs. The laboratory has become a leading test facility for OTEC technology.

India built a one MW floating OTEC pilot plant near Tamil Nadu, and its government continues to sponsor research.

Cycle Types

Cold seawater is an integral part of each of the three types of OTEC systems: closed-cycle, open-cycle, and hybrid. To operate, the cold seawater must be brought to the surface. The primary approaches are active pumping and desalination. Desalinating seawater near the sea floor lowers its density, which causes it to rise to the surface.

The alternative to costly pipes to bring condensing cold water to the surface is to pump vapourised low boiling point fluid into the depths to be condensed, thus reducing pumping volumes and reducing technical and environmental problems and lowering costs.

Closed

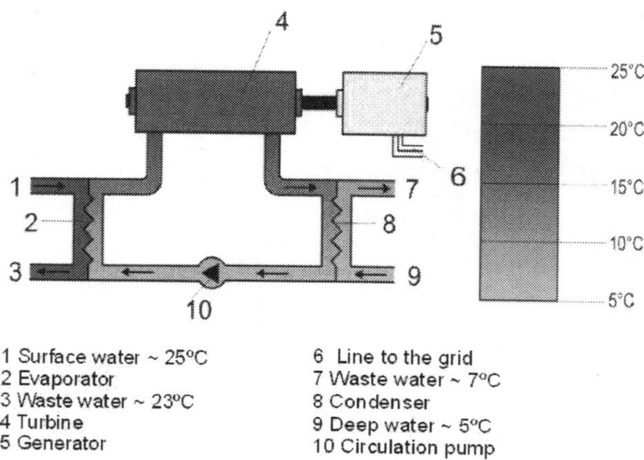

Figure: *Diagram of a closed cycle OTEC plant*

Closed-cycle systems use fluid with a low boiling point, such as ammonia, to power a turbine to generate electricity. Warm surface seawater is pumped through a heat exchanger to vapourise the fluid.

The expanding vapour turns the turbo-generator. Cold water, pumped through a second heat exchanger, condenses the vapour into a liquid, which is then recycled through the system.

In 1979, the Natural Energy Laboratory and several private-sector partners developed the "mini OTEC" experiment, which achieved the first successful at-sea production of net electrical power from closed-cycle OTEC. The mini OTEC vessel was moored 1.5 miles (2.4 km) off the Hawaiian coast and produced enough net electricity to illuminate the ship's light bulbs and run its computers and television.

Open

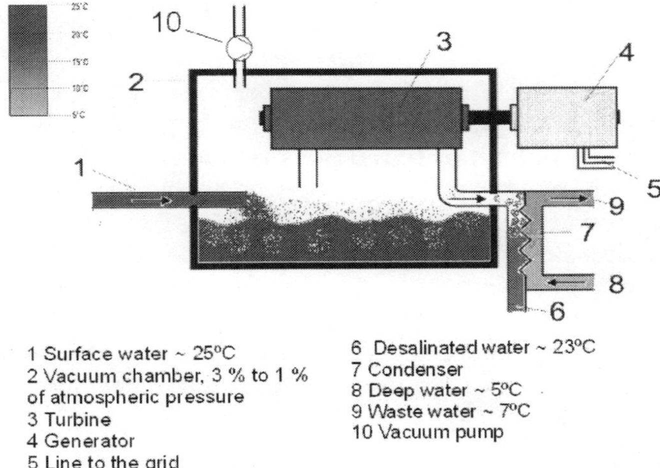

Figure: *Diagram of an open cycle OTEC plant*

Open-cycle OTEC uses warm surface water directly to make electricity. Placing warm seawater in a low-pressure container causes it to boil. The expanding steam drives a low-pressure turbine attached to an electrical generator. The steam, which has left its salt and other contaminants in the low-pressure container, is pure fresh water. It is condensed into a liquid by exposure to cold temperatures from deep-ocean water. This method produces desalinized fresh water, suitable for drinking water or irrigation.

In 1984, the *Solar Energy Research Institute* (now the National Renewable Energy Laboratory) developed a vertical-spout evaporator to convert warm seawater into low-pressure steam for open-cycle plants. Conversion efficiencies were as high as 97% for seawater-to-steam conversion (overall efficiency using a vertical-spout evaporator would still only be a few per cent). In May 1993, an open-cycle OTEC plant at Keahole Point, Hawaii, produced 50,000 watts of electricity during a net power-producing experiment. This broke the record of 40 kW set by a Japanese system in 1982.

Hybrid

A hybrid cycle combines the features of the closed- and open-cycle systems. In a hybrid, warm seawater enters a vacuum chamber and is flash-evaporated, similar to the open-cycle evaporation process. The steam vapourises the ammonia working fluid of a closed-cycle loop on the other side of an ammonia vapouriser. The vapourised fluid then drives a turbine to produce electricity. The steam condenses within the heat exchanger and provides desalinated water.

Working Fluids

A popular choice of working fluid is ammonia, which has superior transport properties, easy availability, and low cost. Ammonia, however, is toxic and flammable. Fluorinated carbons such as CFCs and HCFCs are not toxic or flammable, but they contribute to ozone layer depletion. Hydrocarbons too are good candidates, but they are highly flammable; in addition, this would create competition for use of them directly as fuels. The power plant size is dependent upon the vapour pressure of the working fluid. With increasing vapour pressure, the size of the turbine and heat exchangers decreases while the wall thickness of the pipe and heat exchangers increase to endure high pressure especially on the evaporator side.

Land, Shelf and Floating Sites

Figure: *Left: Pipes used for OTEC. Right: Floating OTEC plant constructed inIndia in 2000*

OTEC has the potential to produce gigawatts of electrical power, and in conjunction with electrolysis, could produce enough hydrogen to completely replace all projected global fossil fuel consumption.

Reducing costs remains an unsolved challenge, however. OTEC plants require a long, large diametre intake pipe, which is submerged a kilometre or more into the ocean's depths, to bring cold water to the surface.

Land-based

Land-based and near-shore facilities offer three main advantages over those located in deep water. Plants constructed on or near land do not require sophisticated mooring, lengthy power cables, or the more extensive maintenance associated with open-ocean environments. They can be installed in sheltered areas so that they are relatively safe from storms and heavy seas. Electricity, desalinated water, and cold, nutrient-rich seawater could be transmitted from near-shore facilities via trestle bridges or causeways. In addition, land-based or near-shore sites allow plants to operate with related industries such as mariculture or those that require desalinated water.

Favoured locations include those with narrow shelves (volcanic islands), steep (15-20 degrees) offshore slopes, and relatively smooth sea floors. These sites minimise the length of the intake pipe. A land-based plant could be built well inland from the shore, offering more protection from storms, or on the beach, where the pipes would be shorter. In either case, easy access for construction and operation helps lower costs.

Land-based or near-shore sites can also support mariculture. Tanks or lagoons built on shore allow workers to monitor and control miniature marine environments. Mariculture products can be delivered to market via standard transport. One disadvantage of land-based facilities arises from the turbulent wave action in the surf zone. Unless the OTEC plant's water supply and discharge pipes are buried in protective trenches, they will be subject to extreme stress during storms and prolonged periods of heavy seas. Also, the mixed discharge of cold and warm seawater may need to be carried several hundred metres offshore to reach the proper depth before it is released. This arrangement requires additional expense in construction and maintenance.

OTEC systems can avoid some of the problems and expenses of operating in a surf zone if they are built just offshore in waters ranging from 10 to 30 metres deep (Ocean Thermal Corporation 1984). This type of plant would use shorter (and therefore less costly) intake and discharge pipes, which would avoid the dangers of turbulent surf. The plant itself, however, would require protection from the marine environment, such as breakwaters and erosion-resistant foundations, and the plant output would need to be transmitted to shore.

Shelf-based

To avoid the turbulent surf zone as well as to move closer to the cold-water resource, OTEC plants can be mounted to the continental shelf at depths up to 100 metres (330 ft). A shelf-mounted plant could be towed to the site and affixed to the sea bottom. This type of construction is already used for offshore oil rigs. The complexities of operating an OTEC plant in deeper water may make them more expensive than land-based approaches. Problems include the stress of open-ocean conditions and more difficult product delivery. Addressing strong ocean currents and large waves adds engineering and construction expense. Platforms require extensive pilings to maintain a stable base. Power delivery can require long underwater cables to reach land. For these reasons, shelf-mounted plants are less attractive.

Floating

Floating OTEC facilities operate off-shore. Although potentially optimal for large systems, floating facilities present several difficulties. The difficulty of mooring plants in very deep water complicates power delivery. Cables attached to floating platforms are more susceptible to damage, especially during storms. Cables at depths greater than 1000 metres are difficult to maintain and repair. Riser cables, which connect the sea bed and the plant, need to be constructed to resist entanglement.

As with shelf-mounted plants, floating plants need a stable base for continuous operation. Major storms and heavy seas can break the vertically suspended cold-water pipe and interrupt warm water intake as well. To help prevent these problems, pipes can be made of flexible polyethylene attached to the bottom of the platform and gimballed with joints or collars. Pipes may need to be uncoupled from the plant to prevent storm damage. As an alternative to a warm-water pipe, surface water can be drawn directly into the platform; however, it is necessary to prevent the intake flow from being damaged or interrupted during violent motions caused by heavy seas.

Connecting a floating plant to power delivery cables requires the plant to remain relatively stationary. Mooring is an acceptable method, but current mooring technology is limited to depths of about 2,000 metres (6,600 ft). Even at shallower depths, the cost of mooring may be prohibitive.

Some Proposed Projects

OTEC projects under consideration include a small plant for the U.S. Navy base on the British overseas territory island of Diego Garcia in the Indian Ocean. OCEES International, Inc. is working with the U.S. Navy on a design for a proposed 13-MW OTEC plant, to replace the current diesel generators. The OTEC plant would also provide 1.25 million gallons per day (MGD) of potable water. A private U.S. company has proposed building a 10-MW OTEC plant on Guam.

Hawaii

Lockheed Martin's Alternative Energy Development team has partnered with Makai Ocean Engineering to complete the final design phase of a 10-MW closed cycle OTEC pilot system which will become operational in Hawaii in the 2012-2013 time frame.

This system is being designed to expand to 100-MW commercial systems in the near future.

In November, 2010 the U.S. Naval Facilities Engineering Command (NAVFAC) awarded Lockheed Martin a US$4.4 million contract modification to develop critical system components and designs for the plant, adding to the 2009 $8.1 million contract and two Department of Energy grants totaling $1 million in 2008 and March 2010.

Related Activities

OTEC has uses other than power production.

Air Conditioning

The 41 °F (5 °C) cold seawater made available by an OTEC system creates an opportunity to provide large amounts of cooling to operations near the plant. The water can be used in chilled-water coils to provide air-conditioning for buildings. It is estimated that a pipe 1 foot (0.30 m) in diametre can deliver 4,700 gallons per minute of water. Water at 43 °F (6 °C) could provide more than enough air-conditioning for a large building. Operating 8,000 hours per year in lieu of electrical conditioning selling for 5-10¢ per kilowatt-hour, it would save $200,000-$400,000 in energy bills annually.

The InterContinental Resort and Thalasso-Spa on the island of Bora Bora uses an OTEC system to air-condition its buildings. The system passes seawater through a heat exchanger where it cools freshwater in a closed loop system. This freshwater is then pumped to buildings and directly cools the air.

Chilled-soil Agriculture

OTEC technology supports chilled-soil agriculture. When cold seawater flows through underground pipes, it chills the surrounding soil. The temperature difference between roots in the cool soil and leaves in the warm air allows plants that evolved in temperate climates to be grown in the subtropics. Dr. John P. Craven, Dr. Jack Davidson and Richard Bailey patented this process and demonstrated it at a research facility at the Natural Energy Laboratory of Hawaii Authority (NELHA). The research facility demonstrated that more than 100 different crops can be grown using this system. Many normally could not survive in Hawaii or at Keahole Point.

Aquaculture

Aquaculture is the best-known byproduct, because it reduces the financial and energy costs of pumping large volumes of water from the deep ocean. Deep ocean water contains high concentrations of essential nutrients that are depleted in surface waters due to biological consumption. This "artificial upwelling" mimics the natural upwellings that are responsible for fertilizing and supporting the world's largest marine ecosystems, and the largest densities of life on the planet.

Cold-water delicacies, such as salmon and lobster, thrive in this nutrient-rich, deep, seawater. Microalgae such as *Spirulina*, a health food supplement, also can be cultivated. Deep-ocean water can be combined with surface water to deliver water at an optimal temperature.

Non-native species such as Salmon, lobster, abalone, trout, oysters, and clams can be raised in pools supplied by OTEC-pumped water. This extends the variety of fresh seafood products available for nearby markets. Such low-cost refrigeration can be used to maintain the quality of harvested fish, which deteriorate quickly in warm tropical regions.

Desalination

Desalinated water can be produced in open- or hybrid-cycle plants using surface condensers to turn evaporated seawater into potable water. System analysis indicates that a 2-megawatt plant could produce about 4,300 cubic metres (150,000 cu ft) of desalinated water each day. Another system patented by Richard Bailey creates condensate water by regulating deep ocean water flow through surface condensers correlating with fluctuating dew-point temperatures. This condensation system uses no incremental energy and has no moving parts.

Hydrogen Production

Hydrogen can be produced via electrolysis using OTEC electricity. Generated steam with electrolyte compounds added to improve efficiency is a relatively pure medium for hydrogen production. OTEC can be scaled to generate large quantities of hydrogen. The main challenge is cost relative to other energy sources and fuels.

Mineral Extraction

The ocean contains 57 trace elements in salts and other forms and dissolved in solution. In the past, most economic analyses concluded that mining the ocean for trace elements would be unprofitable, in part because of the energy required to pump the water. Mining generally targets minerals that occur in high concentrations, and can be extracted easily, such as magnesium. With OTEC plants supplying water, the only cost is for extraction. The Japanese investigated the possibility of extracting uranium and found developments in other technologies (especially materials sciences) were improving the prospects.

Political Concerns

Because OTEC facilities are more-or-less stationary surface platforms, their exact location and legal status may be affected by the United Nations Convention on the Law of the Sea treaty (UNCLOS). This treaty grants coastal nations 3-, 12-, and 200-mile (320 km) zones of varying legal authority from land, creating potential conflicts and regulatory barriers. OTEC plants and similar structures would be considered artificial islands under the treaty, giving them no independent legal status. OTEC plants could be perceived as either a threat or potential partner to fisheries or to seabed mining operations controlled by the International Seabed Authority.

Cost and Economics

For OTEC to be viable as a power source, the technology must have tax and subsidy treatment similar to competing energy sources. Because OTEC systems have not yet been widely deployed, cost estimates are uncertain. One study estimates power generation costs as low as US $0.07 per kilowatt-hour, compared with $0.05 - $0.07 for subsidized wind systems.

Beneficial factors that should be taken into account include OTEC's lack of waste products and fuel consumption, the area in which it is available, (often within 20° of the equator) the geopolitical effects of petroleum dependence, compatibility with alternate forms of ocean power such as wave energy, tidal energy and methane hydrates, and supplemental uses for the seawater.

Mathematics of OTEC

Variation of Ocean Temperature with Depth

The total insolation received by the oceans (covering 70% of the earth's surface, with clearness index of 0.5 and average energy retention of 15%) is 5.457 x e^{10} Megajoules/year (MJ/yr) x.7 x .5 x .15 = 2.87 x e^{10} MJ/yr

We can use Lambert's law to quantify the solar energy absorption by water,

$$-\frac{dI(y)}{dy} = \mu I$$

where, y is the depth of water, I is intensity and ì is the absorption coefficient. Solving the above differential equation,

$$I(y) = I_0 \exp(-\mu y)$$

The absorption coefficient ì may range from 0.05 m"1 for very clear fresh water to 0.5 m"1 for very salty water.

Since the intensity falls exponentially with depth y, heat absorption is concentrated at the top layers. Typically in the tropics, surface temperature values are in excess of 25 °C (77 °F), while at 1 kilometre (0.62 mi), the temperature is about 5–10 °C (41–50 °F). The warmer (and hence lighter) waters at the surface means there are no thermal convection currents. Due to the small temperature gradients, heat transfer by conduction is too low to equalize the temperatures. The ocean is thus both a practically infinite heat source and a practically infinite heat sink.

This temperature difference varies with latitude and season, with the maximum in tropical, subtropical and equatorial waters. Hence the tropics are generally the best OTEC locations.

Open/Claude Cycle

In this scheme, warm surface water at around 27 °C (81 °F) enters an evaporator at pressure slightly below the saturation pressures causing it to vapourise.

$$H_1 = H_f$$

Where H_f is enthalpy of liquid water at the inlet temperature, T_1.

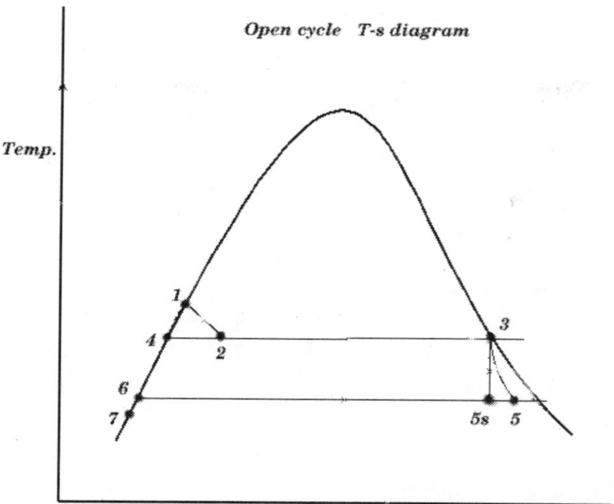

This temporarily superheated water undergoes volume boiling as opposed to pool boiling in conventional boilers where the heating surface is in contact. Thus the water partially flashes to steam with two-phase equilibrium prevailing. Suppose that the pressure inside the evaporator is maintained at the saturation pressure, T_2.

$$H_2 = H_1 = H_f + x_2 H_{fg}$$

Here, x_2 is the fraction of water by mass that vapourises. The warm water mass flow rate per unit turbine mass flow rate is $1/x_2$.

The low pressure in the evaporator is maintained by a vacuum pump that also removes the dissolved non-condensable gases from the evaporator. The evaporator now contains a mixture of water and steam of very low vapour quality (steam content). The steam is separated from the water as saturated vapour. The remaining water is saturated and is discharged to the ocean in the open cycle. The steam is a low pressure/high specific volume working fluid. It expands in a special low pressure turbine.

$$H_3 = H_g$$

Here, H_g corresponds to T_2. For an ideal isentropic (reversible adiabatic) turbine,

$$s_{5,s} = s_3 = s_f + x_{5,s} s_{fg}$$

The above equation corresponds to the temperature at the exhaust of the turbine, T_5. $x_{5,s}$ is the mass fraction of vapour at state 5.

The enthalpy at T_5 is,

$$H_{5,s} = H_f + x_{5,s} H_{fg}$$

This enthalpy is lower. The adiabatic reversible turbine work = H_3-$H_{5,s}$.

Actual turbine work W_T = (H_3-$H_{5,s}$) x *polytropic efficiency*

$$H_5 = H_3 - \text{actual work}$$

The condenser temperature and pressure are lower. Since the turbine exhaust is to be discharged back into the ocean, a direct contact condenser is used to mix the exhaust with cold water, which results in a near-saturated water. That water is now discharged back to the ocean.

H_6=H_f, at T_5. T_7 is the temperature of the exhaust mixed with cold sea water, as the vapour content now is negligible,

$$H_7 \approx H_f \ \ at\ T_7$$

The temperature differences between stages include that between warm surface water and working steam, that between exhaust steam and cooling water, and that between cooling water reaching the condenser and deep water. These represent external irreversibilities that reduce the overall temperature difference.

The cold water flow rate *per* unit turbine mass flow rate,

$$m_c = \frac{H_5 - H_6}{H_6 - H_7}$$

Turbine mass flow rate, $M_T = \dfrac{\text{turbine work required}}{W_T}$

Warm water mass flow rate, $M_w = M_T\, m_w$

Cold water mass flow rate $M_c = M_T m_C$

Closed/Anderson Cycle

Developed starting in the 1960s by J. Hilbert Anderson of Sea Solar Power, Inc. In this cycle, Q_H is the heat transferred in the evaporator from the warm sea water to the working fluid. The working fluid exits the evaporator as a gas near itsdew point.

The high-pressure, high-temperature gas then is expanded in the turbine to yield turbine work, W_T. The working fluid is slightly superheated at the turbine exit and the turbine typically has an efficiency of 90% based on reversible, adiabatic expansion.

From the turbine exit, the working fluid enters the condenser where it rejects heat, $-Q_C$, to the cold sea water. The condensate is then compressed to the highest pressure in the cycle, requiring condensate pump work, W_C. Thus, the Anderson closed cycle is a Rankine-type cycle similar to the conventional power plant steam cycle except that in the Anderson cycle the working fluid is never superheated more than a few degrees Fahrenheit. Owing to viscous effects, working fluid pressure drops in both the evaporator and the condenser. This pressure drop, which depends on the types of heat exchangers used, must be considered in final design calculations but is ignored here to simplify the analysis. Thus, the parasitic condensate pump work, W_C, computed here will be lower than if the heat exchanger pressure drop was included. The major additional parasitic energy requirements in the OTEC plant are the cold water pump work, W_{CT}, and the warm water pump work, W_{HT}. Denoting all other parasitic energy requirements by W_A, the net work from the OTEC plant, W_{NP} is

$$W_{NP} = W_T + W_C + W_{CT} + W_{HT} + W_A$$

The thermodynamic cycle undergone by the working fluid can be analysed without detailed consideration of the parasitic energy requirements. From the first law of thermodynamics, the energy balance for the working fluid as the system is

$$W_N = Q_H + Q_C$$

where $W_N = W_T + W_C$ is the net work for the thermodynamic cycle. For the idealised case in which there is no working fluid pressure drop in the heat exchangers,

$$Q_H = \int_H T_H ds$$

and

$$Q_C = \int_C T_C ds$$

so that the net thermodynamic cycle work becomes

$$W_N = \int_H T_H ds + \int_C T_C ds$$

Subcooled liquid enters the evaporator. Due to the heat exchange with warm sea water, evaporation takes place and usually superheated vapour leaves the evaporator. This vapour drives the turbine and the

2-phase mixture enters the condenser. Usually, the subcooled liquid leaves the condenser and finally, this liquid is pumped to the evaporator completing a cycle.

Technical Difficulties

Dissolved Gases

The performance of direct contact heat exchangers operating at typical OTEC boundary conditions is important to the Claude cycle. Many early Claude cycle designs used a surface condenser since their performance was well understood. However, direct contact condensers offer significant disadvantages. As cold water rises in the intake pipe, the pressure decreases to the point where gas begins to evolve. If a significant amount of gas comes out of solution, placing a gas trap before the direct contact heat exchangers may be justified. Experiments simulating conditions in the warm water intake pipe indicated about 30% of the dissolved gas evolves in the top 8.5 metres (28 ft) of the tube. The trade-off between pre-dearation of the seawater and expulsion of non-condensable gases from the condenser is dependent on the gas evolution dynamics, deaerator efficiency, head loss, vent compressor efficiency and parasitic power. Experimental results indicate vertical spout condensers perform some 30% better than falling jet types.

Microbial Fouling

Because raw seawater must pass through the heat exchanger, care must be taken to maintain good thermal conductivity. Biofouling layers as thin as 25 to 50 micrometres (0.00098 to 0.0020 in) can degrade heat exchanger performance by as much as 50%. A 1977 study in which mock heat exchangers were exposed to seawater for ten weeks concluded that although the level of microbial fouling was low, the thermal conductivity of the system was significantly impaired. The apparent discrepancy between the level of fouling and the heat transfer impairment is the result of a thin layer of water trapped by the microbial growth on the surface of the heat exchanger.

Another study concluded that fouling degrades performance over time, and determined that although regular brushing was able to remove most of the microbial layer, over time a tougher layer formed that could not be removed through simple brushing. The study passed sponge rubber balls through the system. It concluded that although the ball treatment decreased the fouling rate it was not enough to completely halt growth and brushing was occasionally necessary to restore capacity. The microbes regrew more quickly later in the experiment (i.e. brushing became necessary more often) replicating the results of a previous study. The increased growth rate after subsequent cleanings appears to result from selection pressure on the microbial colony.

Continuous use of 1 hour per day and intermittent periods of free fouling and then chlorination periods (again 1 hour per day) were studied. Chlorination slowed but did not stop microbial growth; however chlorination levels of.1 mg per litre for 1 hour per day may prove effective for long term operation of a plant. The study concluded that although microbial fouling was an issue for the warm surface water heat exchanger, the cold water heat exchanger suffered little or no biofouling and only minimal inorganic fouling.

Besides water temperature, microbial fouling also depends on nutrient levels, with growth occurring faster in nutrient rich water. The fouling rate also depends on the material used to construct the heat exchanger. Aluminium tubing slows the growth of microbial life, although the oxide layer which forms on the inside of the pipes complicates cleaning and leads to larger efficiency losses. In contrast, titanium tubing allows biofouling to occur faster but cleaning is more effective than with aluminium.

Sealing

The evaporator, turbine, and condenser operate in partial vacuum ranging from 3% to 1% of atmospheric pressure. The system must be carefully sealed to prevent in-leakage of atmospheric air that can degrade or shut down operation. In closed-cycle OTEC, the specific volume of low-pressure steam is very large compared to that of the pressurised working fluid. Components must have large flow areas to ensure steam velocities do not attain excessively high values.

Parasitic Power Consumption by Exhaust Compressor

An approach for reducing the exhaust compressor parasitic power loss is as follows. After most of the steam has been condensed by spout condensers, the non-condensible gas steam mixture is passed through a counter current region which increases the gas-steam reaction by a factor of five. The result is an 80% reduction in the exhaust pumping power requirements.

Cold Air/Warm Water Conversion

In winter in coastal Arctic locations, seawater can be 40 °C (104 °F) warmer than ambient air temperature. Closed-cycle systems could exploit the air-water temperature difference. Eliminating seawater extraction pipes might make a system based on this concept less expensive than OTEC. This technology is due to H. Barjot, who suggested butane as cryogen, because of its freezing point of ”0.5 °C (31.1 °F) and its non-solubility in water. Assuming a level of efficiency of realistic 4 %, calculations show that the amount of energy generated with one cubic metre water at a temperature of 2 °C (36 °F) in a place with an air temperature of ”22 °C (“8 °F) equals the amount of energy generated by letting this cubic metre water run through a hydroelectric plant of 4000 feet (1,200 m) height.

Barjot Polar Power Plants could be located on island in the polar region or designed as swimming barges or platforms attached to the ice cap. The weather station Myggbuka at Greenlands east coast for example, which is only 2,100 km away from Glasgow, detects monthly mean temperatures below ”15 °C (5 °F) during 6 winter months in the year.

Erie Water Works

The Erie Water Works was incorporated in 1865 as the Erie Water and Gas Company to provide drinking water and fire hydrant water for the city of Erie, Pennsylvania. The Water Works, also known as the Erie City Water Authority, replaced the Erie Water Systems. Its board of commissioners operates independently of the city government.

A simple pump log system provided the first systematic water works for Erie beginning in 1840, tapping spring water on the Reed farm south of 18th Street and Parade Street. The city began inquiries towards a more substantial system beginning in 1853. A state law passed by the Commonwealth of Pennsylvania on 11 March 1857 on the formation and organisation of gas and water companies was the basis for incorporation in 1865 and subsequent construction and provision of utility service.

Infrastructure

Erie's city government decided on 16 July 1866 to seek a hydraulic engineer to oversee construction. The city engaged Henry P M Birkinbine, Esq, of Philadelphia, Pennsylvania, who reported his findings on 23 February 1867. A formal contract to provide fire safety water for twenty years—50 fire hydrants at $9,000 per year—was signed by the city government but never enforced.

The location of the water works at the foot of Chestnut Street was determined in November 1867.

The West Engine Company of Norristown, Pennsylvania was contracted in November 1867 to provide two Cornish Bull engines. A single engine pumped 2.5 million US gallons (9,500 m^3) per day, while 4.0 million US gallons (15,000 m^3) per day were pumped when the engines were doubled. The city contracted with the Holly Manufacturing Company of Lockport, New York in 1885 to provide a Gaskill Engine to provide 5 million US gallons (19,000 m^3) per day of water. That engine was installed in a new pump house on 11 June 1887.

The Erie City Iron Works was contracted in December 1867 to erect a 5-foot-wide (1.5 m) standpipe, which eventually reached a height of 233 feet (71 m), thought to have been the tallest standpipe in the world. John M Kuhn, Esq was hired in 1868 to build the standpipe tower, and Captain James Dunlap was contracted that year to construct the crib work. The foundation for the water works was excavated beginning on 7 April 1868.

The city government decided in 1870 that it needed a reservoir. Seven acres of land at the Cochran estate, south of 26th Street between Chestnut Street and Cherry Street, was acquired for that purpose in 1871. Work on the reservoir was completed in 1874.

The Chestnut Street Water Treatment Plant and Sommerheim Water Treatment Plant filter water received from Lake Erie and the Sommerheim Reservoir, which is spring fed. Sommerheim pumps 25 million US gallons (95,000 m^3) of water per day to homes and businesses in Erie and surrounding communities.

Erie Water Works is the second largest consumer of electricity in Erie after GE Transportation. As a result of a vulnerability assessment conducted by the Department of Homeland Security, the Sommerheim plant is to receive a new power generation facility to back up its Penelec electrical service. The facility, which will cost $1.2 million, is to consist of two diesel-powered electric generators with a day tank and a bulk storage fuel tank.

Organisation

As of 4 April 1867, three water commissioners were appointed by the judge of the Court of Common Pleas to serve staggered three-year terms: Henry Rawle, William Ward Reed (1 April 1824 - 10 January 1904) , and William Lawrence Scott.

The Water Works, also known as the Erie City Water Authority, took over the Erie Water Systems as of 1 January 1992. The Erie

Times-News suggests that the Water Authority replaced the Water System in 1989.

Politics

- The mayoral election of 1867 hinged on water issues, especially whether water should be collected from Presque Isle Bay or from local spring water.
- An effort to stop fluoridation of the water supply was debated in late 2001, but no referendum was added to the November 2001 ballot. Council members threatened to temporarily disband the Erie City Water Authority to remove its board members if they persisted in their plans to fluoridate the water.
- A $21 million development project is aimed at improving water quality at the Sommerheim treatment plant in three phases, with completion expected by early 2009. Filtration is currently handled with a sand and charcoal process, achieving turbidity levels of 0.1 - 0.2 NTU. Replacement of the current system is being weighed against installation of a new sophisticated membrane system at $5 million to $7 million additional cost, adjusting the total cost of the project to as much as $28 million. The membrane system would help the city anticipate new federal water treatment standards, lower pre-treatment costs, and achieve turbidity levels of 0.004 NTU, 125 times better than the state standard of 0.5 NTU.
- Plans have been laid for an extension of water service to the Borough of McKean.

Micro Credit for Water Supply and Sanitation

Micro credit for water supply and sanitation is an innovative application of micro credit to provide loans to small enterprises and households in order to increase access to an improved water source and sanitation in developing countries. While most investments in water supply and sanitation infrastructure are financed by the public sector, current investment levels are insufficient to achieve universal access or, at the minimum, to reach the Millennium Development Goals related to water supply and sanitation. Commercial credit to public utilities is limited by low tariffs and insufficient cost recovery. Micro credits are a complementary or alternative approach to allow the poor to gain access to water supply and sanitation Funding is either provided to small-scale independent water providers who generate an income stream from selling water or to households in order to finance house connections, plumbing installations in their houses, or various forms of on-site sanitation such as septic tanks or latrines. While they have become increasingly common, many microfinance institutions still have only limited experience with financing investments in water supply and sanitation. While there have been many pilot projects in both urban and rural areas, just a small number have actually been scaled up. According to an extensive study of microfinance in water supply and sanitation, there are three types of microcredit products in the water sector across urban and rural areas:

1. Retail loans aiming to improve access to water supply and sanitation at household level,
2. Loans for small and medium enterprises receiving micro credits for small water supply investments and
3. Loans for urban services upgrading and shared facilities in low income areas.

Retail Loans for Household Water and Sanitation

Retail loans for water and sanitation facilities are being offered to individuals by some large and commonly known microfinance institutions such as the Grameen Bank and the Vietnam Bank for Social Policies in South East Asia. In general, these credits are used to finance such things as bathrooms, toilets or water purifiers and range from 30 to 250 USD with a tenure generally less than three years. Although the potential market size is considered as huge in both rural and urban areas and some of these water and sanitation schemes have achieved a significant scale, compared to the microfinance institution's overall size and scope, they still play a minor role.

In 1999, all microfinance institutions in Bangladesh and more recent in Vietnam had reached only about 9 percent and 2.4 percent of rural households respectively. In both countries, the water and sanitation portfolio amounts to less than two percent of the microfinance institution's portfolio in total. However, borrowers for water supply and sanitation comprised 30 percent of total borrowers for Grameen Bank and 10 percent of total borrowers from Vietnam Bank for Social Policies.

Complementary to targeted micro credits, general purpose loans are increasingly gaining weight, as they are used for water and sanitation activities in India and a number of African countries (e.g. Benin, Zambia, and Uganda). For instance, the water and sanitation portfolio of the Indian microfinance institution SEWA Bank comprised 15 percent of all loans provided in the city of Ahmedabad over a period of five years.

Development aid institutions particularly focused on establishing linkage with local microfinance institutions. USAID for instance, has been actively involved by assisting microfinance institutions in designing micro credit products for the water supply sector in Indonesia.

Indonesia: Expanding Customer base Through Cooperation

In 2006, the Bank Rakyat Indonesia signed a Memorandum of Understanding with the water utility PDAM Tanah to initiate water connections schemes on the basis of micro credits creating a win-win situation. The joint venture with technical support of the USAID Environmental Services Program is planned to be extended to other regions. By 2009, the involved institutions intend to have provided 10.000 households with water connections. According to Metha, the effort has resulted in benefits for the utility, the microfinance institution and for the customers who gained access to water supply. The utility benefited from the expanded customer base and additional water sales as it has helped to reduce its average costs by 42 percent over a period of three years and to reduce its non-revenue water ratio from a percentage of 56,5 in 2002 to 36 percent at the end of 2004. Furthermore, the impact assessment report c both institutions conclude, that the financial cooperation could be extended to finance PDAMS's need for additional banking services.

Vietnam: Sanitation Revolving Fund Managed by the Women's Union

In 1999, the World Bank in cooperation with the governments of Australia, Finland and Denmark supported the creation of a Sanitation Revolving Fund with an initial working capital of USD 3 million. The project was carried out in the cities of Danang, Haiphong and Quang Ninh. The overall aim was to provide small loans (USD 145) to low-income and poor households for targeted sanitation investments such as septic tanks, urine diverting/composting latrines or sewer connections. Households willing to participate needed to join a savings and credit group of 12 to 20 people. Members of those groups were required to live near to each other to ensure community control. The loans had a catalyst effect for household investment. With loans covering approximately two thirds of investment costs, households had to find complementary sources of finance (from family and friends). In contrast to a centralised, supply-driven approach, where government institutions design a project with little community consultation and no capacity building for the community, this approach was strictly demand driven and thus required the Sanitation Revolving Fund to develop awareness raising campaigns for sanitation. Managed by the microfinance-experienced Women's Union of Vietnam, the Sanitation Revolving Fund gave 200.000 households the opportunity to finance and build sanitation facilities over a period of seven years. With a leverage effect of up to 25 times the amount of public spending on household investment and repayment rates of almost 100 percent, the fund is seen as a best practice example by its financiers and thus considered to be scaled up with further support of the World Bank and the Vietnam Bank for Social Policies.

Small and Medium Enterprise (SME) Loans for Water and Sanitation

SME-type loans are used for investments by community groups, for private providers in greenfield contexts or for rehabilitation measures of water supply and sanitation. Supplied by mature microfinance institutions, these loans are seen as suitable for other suppliers in the value chain such as pit latrine emptiers and tanker suppliers. With the right frame conditions such as a solid policy environment and clear institutional relationships, there is a market potential for small-scale water supply projects.

In comparison to retail loans on the household level, the experience with loan products for SME is fairly limited. These loan programs remain mostly at pilot level. However, the design of some recent projects using micro credits for community-based service providers in some African countries (such as those of the K-Rep Bank in Kenya and Togo) shows a sustainable expansion potential. In the case of Kenya's K-Rep Bank, the Water and Sanitation Program, which facilitated the project, is already exploring a countrywide scaling up.

Kenya: K-Rep Bank Finances Community Water Projects

Kenya has numerous community-managed small water enterprises. The financial and institutional frame conditions are enhancing the provision of commercial finance. The water and sanitation program in Africa has launched an initiative to use micro credits for the local water and sanitation sector. As part of this initiative, the commercial microfinance bank K-Rep provides loans to 21 community-managed water projects. The Global Partnership on Output-based Aid supports the programme by providing partial subsidies. Every project is pre-financed with a loan up to 80 percent of the project costs (in average USD 80.000). After an independent verification process, certifying a successful completion, a part of the loan is refinanced by a 40 percent Global Partnership on Output-based Aid subsidy. The remaining loan repayments have to be generated from water revenues. In Addition, technical management assistance grants are provided to assist with the project development and to enable the development of market based Business Development Services sector for small water projects. The assistance for further 21 projects have been approved by the Global Partnership on Output-based Aid and the European Union Water Facilities.

Togo: Micro Credits for the Productive use of Rainwater-harvesting Tanks and Shallow Boreholes

In Togo, CREPA (Centre Regional pour l'Eau Potable et L'Assainissement à Faible Côut) was encouraging the liberalisation of water services in 2001. As a consequence, six domestic micro finance institutions were preparing micro credit scheme for a shallow borehole (3000 USD) or rainwater-harvesting tank (1000 USD) for at least two households from a certain area. The loans are originally dedicated to households, which act as small private provider selling water in bulk or in buckets. However the funds are directly disbursed to the private (drilling) companies. In the period from 2001 to 2006, roughly 1200 water points were built and are hence used for small business activities by the households participated in that programme.

Loans for Urban Services Upgrading and Shared Facilities in Low Income Areas

According to UN-Habitat, by 2020 the number people living in the urban slums of developing countries without adequate WSS services will increase to about 1.4 billion. In this scenario, micro credits could be used as an instrument to finance the upgrade or building of shared facilities in "slum areas". Engineering has been an aspect of life since the beginnings of human existence. The earliest practices of Civil engineering may have commenced between 4000 and 2000 BC in Ancient Egypt and Mesopotamia (Ancient Iraq) when humans started to abandon a nomadic existence, thus causing a need for the construction of shelter. During this time, transportation became increasingly important leading to the development of the wheel and sailing.

Until modern times there was no clear distinction between civil engineering and architecture, and the term engineer and architect were mainly geographical variations referring to the same person, often used interchangeably. The construction of Pyramids in Egypt (circa 2700–2500 BC) might be considered the first instances of large structure constructions. Other ancient historic civil engineering constructions include the Qanat water management system (the oldest older than 3000 years and longer than 71 km,) the Parthenon by Iktinos in Ancient Greece (447–438 BC), the Appian Way by Roman engineers (c. 312 BC), the Great Wall of China by General Meng T'ien under orders from Ch'in Emperor Shih Huang Ti (c. 220 BC) and the stupas constructed in ancient Sri Lanka like the Jetavanaramaya and the extensive irrigation works in Anuradhapura. The Romans developed civil structures throughout their empire, including especially aqueducts, insulae, harbours, bridges, dams and roads.

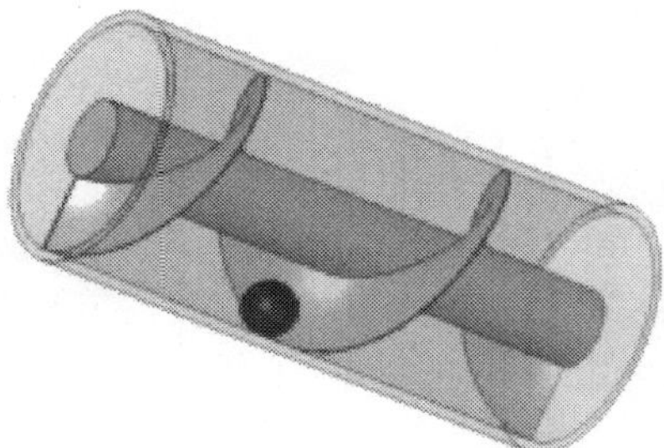

Figure: *The Archimedes screw was operated by hand and could raise water efficiently.*

In the 18th century, the term civil engineering was coined to incorporate all things civilian as opposed to military engineering. The first self-proclaimed civil engineer was John Smeaton who constructed the Eddystone Lighthouse. In 1771 Smeaton and some of his colleagues formed the Smeatonian Society of Civil Engineers, a group of leaders of the profession who met informally over dinner. Though there was evidence of some technical meetings, it was little more than a social society.

In 1818 the Institution of Civil Engineers was founded in London, and in 1820 the eminent engineer Thomas Telford became its first president. The institution received a Royal Charter in 1828, formally recognising civil engineering as a profession. Its charter defined civil engineering as:

The art of directing the great sources of power in nature for the use and convenience of man, as the means of production and of traffic in states, both for external and internal trade, as applied in the construction of roads, bridges, aqueducts, canals, river navigation and docks for internal intercourse and exchange, and in the construction of ports, harbours, moles, breakwaters and lighthouses, and in the art of navigation by artificial power for the purposes of commerce, and in the construction and application of machinery, and in the drainage of cities and towns.

The first private college to teach Civil Engineering in the United States was Norwich University founded in 1819 by Captain Alden Partridge. The first degree in Civil Engineering in the United States was awarded by Rensselaer Polytechnic Institute in 1835. The first such degree to be awarded to a woman was granted by Cornell University to Nora Stanton Blatch in 1905.

Geometric Dimensioning and Tolerancing

Geometric dimensioning and tolerancing (GD&T) is a system for defining and communicating engineering tolerances. It uses a symbolic language on engineering drawings and computer-generated three-dimensional solid models for explicitly describing nominal geometry and its allowable variation. It tells the manufacturing staff and machines what degree of accuracy and precision is needed on each facet of the part.

Overview

Geometric dimensioning and tolerancing (GD&T) is used to define the nominal (theoretically perfect) geometry of parts and assemblies, to define the allowable variation in form and possibly size of individual features, and to define the allowable variation between features. Dimensioning and tolerancing and geometric dimensioning and tolerancing specifications are used as follows:

- Dimensioning specifications define the nominal, as-modelled or as-intended geometry. One example is a basic dimension.
- Tolerancing specifications define the allowable variation for the form and possibly the size of individual features, and the allowable variation in orientation and location between features. Two examples are linear dimensions and feature control frames using a datum reference.

There are several standards available worldwide that describe the symbols and define the rules used in GD&T. One such standard is American Society of Mechanical Engineers (ASME) Y14.5-2009. This article is based on that standard, but other standards, such as those from the International Organisation for Standardisation (ISO), may vary slightly. The Y14.5 standard has the advantage of providing a fairly complete set of standards for GD&T in one document. The ISO standards, in comparison, typically only address a single topic at a time. There are separate standards that provide the details for each of the major symbols and topics below (e.g. position, flatness, profile, etc.).

Dimensioning and Tolerancing Philosophy

According to the ASME Y14.5-2009 standard, the purpose of geometric dimensioning and tolerancing (GD&T) is to describe the engineering intent of parts and assemblies. This is not a completely correct explanation of the purpose of GD&T or dimensioning and tolerancing in general.

The purpose of GD&T is more accurately defined as describing the geometric requirements for part and assembly geometry. Proper application of GD&T will ensure that the allowable part and assembly geometry defined on the drawing leads to parts that have the desired form and fit (within limits) and function as intended.

There are some fundamental rules that need to be applied (these can be found on page 6 of the 2009 edition of the standard):

- All dimensions must have a tolerance. Every feature on every manufactured part is subject to variation, therefore, the limits of allowable variation must be specified. Plus and minus tolerances may be applied directly to dimensions or applied from a general tolerance block or general note. For basic dimensions, geometric tolerances are indirectly applied in a related Feature Control Frame. The only exceptions are for dimensions marked as minimum, maximum, stock or reference.
- Dimensioning and tolerancing shall completely define the nominal geometry and allowable variation. Measurement and scaling of the drawing is not allowed except in certain cases.
- Engineering drawings define the requirements of finished (complete) parts. Every dimension and tolerance required to define the finished part shall be shown on the drawing. If additional dimensions would be helpful, but are not required, they may be marked as reference.
- Dimensions should be applied to features and arranged in such a way as to represent the function of the features.
- Descriptions of manufacturing methods should be avoided. The geometry should be described without explicitly defining the method of manufacture.
- If certain sizes are required during manufacturing but are not required in the final geometry (due to shrinkage or other causes) they should be marked as non-mandatory.

- All dimensioning and tolerancing should be arranged for maximum readability and should be applied to visible lines in true profiles.
- When geometry is normally controlled by gage sizes or by code (e.g. stock materials), the dimension(s) shall be included with the gage or code number in parentheses following or below the dimension.
- Angles of 90° are assumed when lines (including centre lines) are shown at right angles, but no angular dimension is explicitly shown. (This also applies to other orthogonal angles of 0°, 180°, 270°, etc.)
- Dimensions and tolerances are valid at 20 °C / 101.3 kPa unless stated otherwise.
- Unless explicitly stated, all dimensions and tolerances are only valid when the item is in a free state.
- Dimensions and tolerances apply to the full length, width, and depth of a feature including form variation.
- Dimensions and tolerances only apply at the level of the drawing where they are specified. It is not mandatory that they apply at other drawing levels, unless the specifications are repeated on the higher level drawing(s).

(Note: The rules above are not the exact rules stated in the ASME Y14.5-2009 standard.)

GD&T Data Exchange

Exchange of geometric dimensioning and tolerancing (GD&T) information between CAD systems is available on different levels of fidelity for different purposes:

- In the early days of CAD exchange only lines, texts and symbols were written into the exchange file. A receiving system could display them on the screen or print them out, but only a human could interpret them.
- *GD&T presentation*: On a next higher level the presentation information is enhanced by grouping them together into *callouts* for a particular purpose, e.g. a *datum feature callout* and a *datum reference frame*. And there is also the information which of the curves in the exchange file are leader, projection or dimension curves and which are used to form the shape of a product.

- *GD&T representation*: Unlike GD&T presentation, the GD&T representation does not deal with how the information is presented to the user but only deal with which element of a shape of a product has which GD&T characteristic. A system supporting GD&T representation may display the GD&T information in some tree and other dialogs and allow the user to directly select and highlight the corresponding feature on the shape of the product, 2D and 3D.
- Ideally both GD&T presentation and representation are available in the exchange file and are associated with each other. Then a receiving system can allow a user to select a GD&T callout and get the corresponding feature highlighted on the shape of the product.
- An enhancement of GD&T representation is defining a formal language for GD&T (similar like a programming language) which also has built-in rules and restrictions for the proper GD&T usage. This is still a research area.
- *GD&T validation*: Based on GD&T representation data (but not on GD&T presentation) and the shape of a product in some useful format (e.g. a boundary representation), it is possible to validate the completeness and consistency of the GD&T information. The software tool FBTol from the Kansas City Plant is probably the first one in this area.
- GD&T representation information can also be used for the software assisted manufacturing planning and cost calculation of parts.

Engineer's Scale

An engineer's scale is a tool for measuring distances and transferring measurements at a fixed ratio of length. It is commonly made of plastic or aluminium and is just over 12 inches (305 mm) long, but with only 12 inches of markings, leaving the ends unmarked so that the first and last measuring ticks do not wear off. It is used in making engineering drawings, commonly called blueprints, blue lines or plans in a specific scale. For example, "one-tenth size" would appear on a drawing to indicate a part larger than the drawing on the paper itself. It is not to be used to measure machined parts to see if they meet specifications.

In scientific and engineering terminology, a device to measure linear distance and create proportional linear measurements is called a scale. A device for drawing straight lines is a straight edge or ruler. In common usage both are referred to as a ruler.

Figure: *Boxed set of 1850s ivory engineer's scales presented to the railway civil engineer George Turnbull in India. 16 scales are engraved.*

In Canada and the United States, this scale is divided into decimalized fractions of an inch, but has a cross-section like an equilateral triangle, which enables the scale to have six edges indexed for measurement. One edge is divided into tenths of an inch, and the subsequent ones are directly marked for twentieths, thirtieths, fortieths, fiftieths, and finally sixtieths of an inch. Referred to as 1:10, 1:20, 1:30,1:40, 1:50 or 1:60 scale. Typically in civil engineering applications, 1:10 (1"=10') is used exclusively for detail drawings. 1:20 and 1:40 scales are used for working plans. 1:60 is normally used only to show large areas of a project.

The engineer's scale came into existence when machining parts required a greater precision than the usual, binary fractionalization of the inch as in the architect's scale for houses and furniture. They were used, for example, in laying out printed circuit boards with the spacing of leads from integrated circuits as one-tenth of an inch.

On the other hand, they may have come into existence far earlier for use in civil engineering applications when project scope could far exceed the potential of an architect's scale.

The system of imperial units or the imperial system is the system of units first defined in the British Weights and Measures Act of 1824, which was later refined (until 1959) and reduced. The system came into official use across the British Empire. By the late 20th century, most nations of the former empire had officially adopted the metric system as their main system of measurement, although as of 2011 the UK had only partially adopted it. Since 1959, the US and the British yard have been defined identically to be 0.9144 metres, to match the *international yard*. Metric equivalents in this article usually assume this latest official definition. Before this date, the most precise measurement of the Imperial Standard Yard was 0.914398416 metres.

Volume

In 1824, the United Kingdom adopted a close approximation to the ale gallon known as the imperial gallon. The imperial gallon was based on the volume of 10 lb of distilled water weighed in air with brass weights with the barometre standing at 30 in Hg at a temperature of 62 °F.

In 1963 this definition was refined as the space occupied by 10 lb of distilled water of density 0.998859 g/mL weighed in air of density 0.001217 g/mL against weights of density 8.136 g/mL. This works out to 4.546096 L, or 277.4198 cu in. The Weights and Measures Act of 1985 switched to a gallon of exactly 4.54609 L (approximately 277.4194 cu in).

British Apothecaries' Volume Measures

These measurements were in use from 1824, when the new imperial gallon was defined, but fell out of use long before they were officially abolished in 1971.

Table : *Table of British apothecaries' volume units*

Unit	*Previous Unit*	*Metric Value*
minim	...	59.1938802 μl
fluid scruple	20 minims	1.1838776 ml
fluid drachm	3 fluid scruples	3.5516328 ml
fluid ounce	8 fluid drachms	28.4130625 ml
pint	20 fluid ounces	568.26125 ml
gallon	8 pints	4.54609 l

Mass

In the 19th and 20th centuries the UK used three different systems for mass and weight:

- troy weight, used for precious metals;
- avoirdupois weight, used for most other purposes; and
- apothecaries' weight, now virtually unused since the metric system is used for all scientific purposes.

The troy pound (373.2417216 g) was made the primary unit of mass by the 1824 Act; however, its use was abolished in the UK on 6 January 1879, making the Avoirdupois pound the primary unit of mass with only the troy ounce (31.1034768 g) and its decimal subdivisions retained. In all the systems, the fundamental unit is the pound, and all other units are defined as fractions or multiples of it.

Relation to other Systems

The imperial system is one of many systems of English or foot-pound-second units, so named because of the base units of length, mass and time. Although most of the units are defined in more than one system, some subsidiary units were used to a much greater extent, or for different purposes, in one area rather than the other. The distinctions between these systems are often not drawn precisely.

One such distinction is that between these systems and older British/English units/systems or newer additions. The term *imperial* should not be applied to English units that were outlawed in the Weights and Measures Act 1824 or earlier, or which had fallen out of use by that time, nor to post-imperial inventions such as the slug or poundal.

United States Customary Units

The US customary system, which is historically derived from the English units which were in use at the time of settlement. Because the United States was already independent at the time, these units were unaffected by the introduction of the imperial system. Units of length and area are mostly shared between the imperial and US systems, albeit being partially and temporarily defined differently. Capacity measures differ the most due to the introduction of the imperial gallon and the unification of wet and dry measures. The avoirdupois system applies only to weights; it has a *long* designation and a *short* designation for the hundredweight and ton.

Current use of Imperial Units

United Kingdom

British law now defines each imperial unit in terms of the metric equivalent. The metric system is in official use within the United Kingdom for some applications; however, use of Imperial unit is widespread in many cases.

The Units of Measurement Regulations 1995 require that all measuring devices used in trade or retail be capable of measuring and displaying metric quantities. This has now been proven in court against the so-called "Metric Martyrs", a small group of market traders who insisted on trading in imperial units only. Contrary to the impression given by some press reports, these regulations do not currently place any obstacle in the way of using imperial units alongside metric units. Almost all traders in the UK will accept requests from customers specified in imperial units, and scales which display in both unit systems are commonplace in the retail trade. Metric price signs may be accompanied by imperial price signs (known as supplementary indicators) provided that the imperial signs are no larger and no more prominent than the official metric ones. The EU units of measurement directive (directive 80/181/EEC) had previously permitted the use of *supplementary indicators* (imperial measurements) until 31 December 2009, but a revision of the directive published on 11 March 2009 permitted their use indefinitely.

The United Kingdom completed its legal partial transition to the metric system (sometimes referred to as "SI" from the French Système International d'Unités) in 1995, with many imperial units still legally mandated for some application; draught beer *must* be sold in pints, road-sign distances *must* be in yards and miles, length and width (but not weight) restrictions *must* be in feet and inches on road signs, and road speed limits *must* be in miles per hour, therefore instruments in vehicles sold in the UK must be capable of displaying miles per hour. Foreign vehicles, such as all post-2005 Irish vehicles, may legally have instruments displayed only in kilometres per hour. Even though the troy pound was outlawed in the UK in the Weights and Measures Act of 1878, the *troy ounce* still *may* be used for the weight of precious stones and metals. The railways are also a big user of imperial units, with distances officially measured in miles and yards or miles and chains, and also feet and inches, and speeds are in miles per hour, although many modern metro and tram systems are entirely metric, and London Underground uses both metric (for distances) and imperial (for speeds). Metric is also used for the Channel Tunnel and on High Speed 1. Adjacent to Ashford International railway station and Dollands Moor Freight Yard, railway speeds are given in both metric and imperial units.

The use of SI units is required by law for the retail sale of some foods and other commodities, whilst imperial units are required for others. Many British people still use imperial units in everyday life for distance (yards, feet and inches), weight (especially stones and pounds) and volume (gallons & pints). Milk is available in both litre and pint based containers. Most people still measure their weight in stone and pounds, and height in feet and inches (but these must be converted to metric if recorded officially, for example in medical records). Petrol is often quoted as being so much per gallon, despite having been sold exclusively in litres for two decades. Likewise, fuel consumption for cars is still almost always in miles per gallon, though official figures always include litres per 100 km equivalents. Fahrenheit equivalents are occasionally given after Celsius in weather forecasts, though this is becoming rare. Threads on non-metric nuts and bolts etc., are sometimes referred to as Imperial, especially in the UK.

Canada

In 1973 the metric system and SI units were introduced in Canada to replace the imperial system. Within the government, efforts to implement the metric system were extensive; almost any agency, institution, or function provided by the government uses SI units exclusively. Imperial units were eliminated from all road signs, although both systems of measurement will still be found on privately owned signs, such as the height warnings at the entrance of a multi-storey parking facility. In the 1980s, momentum to fully convert to the metric system stalled when the government of Brian Mulroney was elected. There was heavy opposition to metrication and as a compromise the government maintains legal definitions for and allows use of imperial units as long as metric units are shown as well. The law requires that measured products (such as fuel and meat) be priced in metric units, although an imperial price can be shown if a metric price is present. However, there tends to be leniency in regards to fruits and vegetables being priced in imperial units only. However unlike in the rest of Canada, metrication in the Francophone province of Quebec has been more implemented and metric measures are more consistently used in Quebec than elsewhere in Canada both officially and among the population.

Environment Canada still offers an imperial unit option beside metric units, even though weather is typically measured and reported in metric units in the Canadian media. However, some radio stations near the United States border (such as CIMX and CIDR) primarily use imperial units to report the weather. Railways in Canada also continue to use Imperial units.

Imperial units are still used in ordinary conversation. Many Canadians use Imperial units to describe their weight and height; newborns are measured in SI at hospitals, but the birth weight and length is usually announced to family and friends in imperial units. Drivers' licences use SI units. In livestock auction markets, cattle are sold in dollars per hundredweight (short), whereas hogs are sold in dollars per hundred kilograms. Imperial units still dominate in recipes, construction, house renovation and gardening. Land is now surveyed and registered in metric units, although initial surveys used imperial units. For example, partitioning of farm land on the prairies in the late 19th and early 20th centuries was done in imperial units; this accounts for imperial units of distance and area retaining wide use in the Prairie Provinces. The size of most apartments, condominiums and houses continues to be described in square feet rather than square metres, and carpet or flooring tile is purchased by the square foot. Motor-vehicle fuel consumption is reported in both litres per 100 km and statute miles per imperial gallon, leading to the erroneous impression that Canadian vehicles are 20% more fuel-efficient than their apparently identical American counterparts for which fuel economy is reported in statute miles per US gallon (Neither country specifies which gallon is used). Canadian railways maintain exclusive use of imperial measurements to describe train length (feet), train height (feet), capacity (tons), speed (mph), and trackage (miles).

Imperial units also retain common use in firearms and ammunition. Imperial measures are still used in the description of cartridge types, even when the cartridge is of relatively recent invention (e.g., 0.204 Ruger, 0.17 HMR, where the calibre is expressed in decimal fractions of an inch). However, ammunition which is classified in metric already is still kept metric (e.g., 9 mm). In the manufacture of ammunition, bullet and powder weights are expressed in terms of grains for both metric and imperial cartridges. As in most of the western world, air navigation is based on *nautical* units, e.g., the nautical mile, which is neither imperial nor metric.

Australia

In Australia, imperial measurements are still encountered peripherally in either spoken or written form.

Rural land areas are sometimes given in acres. Australian beer glass sizes are based on older imperial sizes but rounded to the nearest 5 ml, while some surf reports are given in feet.

Republic of Ireland

The Republic of Ireland has officially changed over to the metric system since entering the European Union, with distances on new road signs being metric since 1977 and speed limits being metric since 2005. The imperial system remains in limited use - for sales of beer in pubs (traditionally sold by the pint). All other goods are required by law to be sold in metric units, although old quantities are retained for some goods like butter, which is sold in 454-gram (1 lb) packaging. The majority of cars sold pre-2005 feature speedometres with miles per hour. The imperial system is still often used in everyday conversation, especially in the terms of height and weight, particularly by the older generation.

Other Countries

Some imperial measurements remain in limited use in India, Malaysia and Hong Kong. Real estate agents continue to use acres and square feet to describe area in conjunction with hectares and square metres. Measurements in feet and inches, especially for a person's height, are frequently met in conversation and non-governmental publications. In India, inches, feet, yards and degrees Fahrenheit are often used in conjunction with their metric counterparts, while area is measured in acres exclusively (hectares are only used in government documents).

Towns and villages in Malaysia with no proper names had adopted the Malay word *batu* (meaning "rock") to indicate their locations along a main road before the use of metric system (for example, *batu enam* means "6th mile" or "mile 6"). Many of their names remain unchanged even after the adoption of the metric system for distance in the country.

Petrol is still sold by the imperial gallon in Anguilla, Antigua and Barbuda, Belize, Burma, the Cayman Islands, Ecuador, Grenada, Guyana, Sierra Leone and the United Arab Emirates. The United Arab Emirates Cabinet in 2009 issued the Decree No. (270 / 3) specifying that from 1 January 2010 the new unit sale price for petrol will be the litre and not the gallon. This in line with the UAE Cabinet Decision No. 31 of 2006 on the national system of measurement, which mandates the use of International System of units (SI) as a basis for the legal units of measurement in the country. Sierra Leone switched to selling fuel by the litre in May 2011. Antigua and Barbuda will convert to litres before 2015.

Metric System

The metric system is an international decimalised system of measurement. France was first to adopt it in 1799 and it is now the official system of measurement used in almost every country in the world, the United States being the only industrialised country in the world that uses customary units (based on English units) as its official system of measurement, although the metric system has been sanctioned for use there since 1866. Although the United Kingdom has officially adopted the metric system for many measurement applications, it is still not in universal use there and the customary imperial system is still in common and widespread use. Although the originators intended to devise a system that was equally accessible to all, it proved necessary to use prototype units under the custody of government or other approved authorities as standards. Until 1875, control of the prototype units of measure was maintained by the French Government when it passed to an inter-governmental organisation – the Conférence générale des poids et mesures (CGPM). It is now hoped that the last of these prototypes can be retired by 2015.

From its beginning, the main feature of the metric system was the standard set of inter-related base units and a standard set of prefixes in powers of ten. These base units are used to derive larger and smaller units and replaced a huge number of unstandardised units of measure that existed previously. While the system was first developed for commercial use, its coherent set of units made it particularly suitable for scientific and engineering purposes.

The uncoordinated use of the metric system by different scientific and engineering disciplines, particularly in the late 19th century, resulted in different choices of fundamental units, even though all were based on the same definitions of the metre and the kilogram. During the 20th century, efforts were made to rationalise these units and in 1960 the CGPM published the International System of Units ("*Système international d'unités*" in French, hence "SI") which, since then, has been the internationally recognised standard metric system.

Although the metric system has changed and developed since its inception, its basic features have remained constant.

Universality

The metric system was, in the words of the French philosopher Condorcet to be "for all people for all time". It was designed for ordinary people, for engineers who worked in human-related measurements and for astronomers and physicists who worked with numbers both small and large, hence the huge range of prefixes that have now been defined in SI.

The metric system was designed to be universal, that is, available to all. When the French Government first investigated the idea of overhauling their system of measurement, Talleyrand, in the late 1780s, acting on Concordet's advice, invited Riggs, a British Parliamentarian and Jefferson, the American Secretary of State to George Washington, to work with the French in producing an international standard by promoting legislation in their respective legislative bodies. However, these overtures failed and the custody of the metric system remained in the hands of the French Government until 1875.

Decimal Multiples

The metric system is decimal. All multiples and divisions of the base units are factors of the power of ten, an idea first proposed by Stevin in 1586. Thus all lengths and distances, for example, are measured in metres, millimetres ($^1D_{1000}$ of a metre), or kilometres (1000 m). There is no profusion of different units with different conversion factors, such as inches, feet, yards, fathoms, rods, chains, furlongs, miles, nautical miles, leagues, etc. The practical benefits of a decimal system have also been used to replace other non-decimal systems such as currencies, with decimal systems. The simple decimal prefixes encouraged the adoption of the metric system; the use of base 10 arithmetic aids in unit conversion, as compared to 12 inches per foot, 3 feet per yard, et cetera. Differences in expressing units are simply a matter of shifting the decimal point or changing an exponent; for example, the speed of light may be expressed as 299792.458 km/s or 2.99792458×10^8 m/s.

Prefixes

All derived units use a common set of prefixes for each multiple. (This idea was first suggested by Mouton in 1670.) Thus the prefix *kilo* is used for mass (*kilogram*) or length (*kilometre*) both indicating a thousand times the base unit. This did not prevent the popular use of names for some derived units such as the tonne which is a *megagram*; derived from old customary units and rounded to metric.

The function of the prefix is to multiply or divide the measure by a factor of ten, one hundred or a positive integer power of one thousand. Initially positive powers of ten had Greek-derived prefixes and negative power of ten Latin-derived prefixes. However later SI extensions to the prefix system did not follow the Greek-greater-than-one/Latin-less-than-one convention. The most familiar ones from everyday use are the *kilo-*, which is of Greek origin, and the *centi-* and *milli-* which are of Latin origin. Most other metric measurements also use prefixes rooted in Latin or Greek: *deci-* has a Latin root, while *nano-*, *micro-*, *deca-*, *hecto-*, *mega-*, *giga-* have Greek roots. During the 19th century, the factor ten thousand was also used. The corresponding prefixes *myria~* 10^4 and *myrio~* 10^{-4} were both Greek-derived.

Replicable Prototypes

The initial way to establish a standard was to make prototypes of the base units and distribute copies to approved centres. This made the new standard reliant on the original prototypes, which would be in conflict with the previous goal, since all countries would have to refer to the one holding the prototypes.

Instead, where possible, definitions of the base units were developed so that any laboratory equipped with proper instruments would be able to construct their own copies of the standard. In the original version of the metric system the base units could be derived from a specified length (the metre) and the weight [mass] of a specified volume ($^{1}D_{1000}$ of a cubic metre) of pure water. Initially the Assemblée Constituante considered defining the metre as the length of a pendulum that has a period of one second at 45°N and an altitude equal to sea level. The altitude and latitude were specified to accommodate variations in gravity – the specified latitude was a compromise between the latitude of London (51° 30'N), Paris (48° 50'N) and the median parallel of the United States (38°N) to accommodate variations. However Borda persuaded the Assemblée Constituante that a survey having its ends at sea level and based on a meridian that spanned at least 10% of the earth's quadrant would be more appropriate for such a basis.

Realisability

The base units used in the metric system must be realisable, ideally with reference to natural phenomena rather than unique artifacts. Each of the base units in SI is accompanied by a *mise en pratique* published by the BIPM that describes in detail at least one way in which the base unit can be measured.

Two of the base units originally depended on artifacts – the metre and the kilogram. The original prototypes of each artifact were adopted in 1799 and replaced in 1889. The 1889 prototypes used the best technology of the day to ensure stability.

In 1889 there was no generally accepted theory regarding the nature of light but by 1960 the wavelength of specific light spectra could give a more accurate and reproducible value than a prototype metre. In that year the prototype metre was replaced by a formal definition which defines the metre in terms of the wavelength of specified light spectra.

By 1983 it was accepted that the speed of light was constant and that this provided a more reproducible procedure for measuring length, the metre was redefined in terms of the speed of light. These definitions give a much better reproducibility and also allow anyone, anywhere to make their own "standard" metre (assuming that they have a good enough laboratory).

Figure: *K48, above, came from the second batch of kilogram replicas to be produced. It was delivered to Denmark in **1949** with an official mass of 1 kg?+?81 μg. Like all other replicas, it is stored under two nested bell jars virtually all the time.*

Similarly, at the 13th convocation of the GCPM in 1968, the definitive second was redefined in terms of measurements taken from atomic clocks rather than from the earth's rotation; in 2008 the solar day was 0.002 s longer than in 1820 but it was only in the 1960s that this could be measured more accurately using clocks rather than relying on astronomy.

Coherence

Each variant of the metric system has a degree of coherence – the various derived units being directly related to the base units without the need of intermediate conversion factors. Whereas the cgs system had two units of energy, the erg that was related to mechanics and the calorie that was related to thermal energy, coherence was a design aim of SI resulting in only one unit of energy – the joule. For example, the units of force, energy and power are chosen so that the equations

$$\text{force} = \text{mass} \times \text{acceleration}$$

$$\text{energy} = \text{force} \times \text{distance}$$

$$\text{energy} = \text{power} \times \text{time}$$

Hold without the introduction of constant factors. Many relationships in physics, including Einstein's mass-energy equation,

$E = mc^2$, do not require extraneous constants when expressed in coherent units.

In SI, which is a coherent system, the unit of power is the "watt" which is defined as "one joule per second". In the foot-pound-second system of measurement, which is non-coherent, the unit of power is the "horsepower" which is defined as "550 foot-pounds per second", the pound in this context being the pound-force.

Other defined units are derived in a similar way building up on the base units.

History

In 1586, the Flemish mathematician Simon Stevin published a small pamphlet called *De Thiende* ("the tenth"). Decimal fractions had been employed for the extraction of square roots some five centuries before his time, but nobody used decimal numbers in daily life. Stevin declared that using decimals was so important that the universal introduction of decimal weights, measures and coinage was only a matter of time. The idea of a metric system was proposed by John Wilkins, first secretary of the Royal Society of London in 1668. Two years later, in 1670, Gabriel Mouton, a French abbot and scientist, proposed a decimal system of measurement based on the circumference of the Earth. His suggestion was that a unit, the milliare, be defined as a minute of arc along a meridian. He then suggested a system of sub-units, dividing successively by factors of ten into the centuria, decuria, virga, virgula, decima, centesima, and millesima. His ideas attracted interest at the time, and were supported by both Jean Picard and Christiaan Huygens in 1673, and also studied at the Royal Society in London. In the same year, Gottfried Leibniz independently made proposals similar to those of Mouton.

In pre-revolutionary Europe, each state had its own system of units of measure. Some countries, such as Spain and Russia, saw the advantages of harmonising their units of measure with those of their trading partners. However, vested interests who profited from variations in units of measure opposed this. This was particularly prevalent in France where there was a huge inconsistency in the size of units of measure. During the early years of the French Revolution, the leaders of the French revolutionary Assemblée Constituante decided that rather than standardising the size of the existing units, they would a introduce a completely new system based on the principles of logic and natural phenomena.

Initially France attempted to work with other countries towards the adoption of a common set of units of measure. Among the supporters of such an international system of units was Thomas Jefferson who, in 1790, presented a document *Plan for Establishing Uniformity in the Coinage, Weights, and Measures of the United States* to congress in which he advocated a decimal system that used traditional names for units (such as ten inches per foot). The report was considered but not adopted by Congress. There was little support from other countries.

Original System

The law of 18 Germinal, Year III (7 April 1795) defined five units of measure:

- The metre for length
- The are (100 m^2) for area [of land]
- The stère (1 m^3) for volume of stacked firewood
- The litre (1 dm^3) for volumes of liquid
- The gram for mass.

This system continued the tradition of having separate base units for geometrically related dimensions, e.g., metre for lengths, are (100 m^2) for areas, stère (1 m^3) for dry capacities, and litre (1 dm^3) for liquid capacities. The hectare, equal to a hundred ares, is the area of a square 100 metres on a side (about 2.47 acres), and is still in use. The early metric system included only a few prefixes from milli (one thousandth) to *myria* (ten thousand), and they were based on powers of 10 unlike later prefixes added in the SI, which are based on powers of 1,000.

Originally the kilogram was called the "grave"; the "gram" being an alternative name for a thousandth of a grave. However, the word "grave", being a synonym for the title "count" had aristocratic connotations and was renamed the kilogram. France officially adopted the metric system on 10 December 1799 with conversion being mandatory first in Paris and then across the provinces.

International Adoption

The areas that were annexed by France during the Napoleonic era inherited the metric system. In 1812, Napoleon introduced a system known as *mesures usuelles* which used the names of pre-metric units of measure, but defined them in terms of metric units – for example, the *livre metrique* (metric pound) was 500 g and the *toise metrique* (metric fathom) was 2 metres. After the Congress of Vienna in 1815, France lost the territories that she had annexed; some, such as the Papal States reverted to their pre-revolutionary units of measure, others such as Baden adopted a modified version of the *mesures usuelles*, but France kept her system of measurement intact.

In 1817, the Netherlands reintroduced the metric system, but used pre-revolutionary names – for example 1 cm became the *duim* (thumb), the *ons* (ounce) became 100 g and so on. Certain German states adopted similar systems and in 1852 the German Zollverein (customs union) adopted the zollpfund (customs pound) of 500 g for intrastate commerce.

In 1872 the newly formed German Empire adopted the metric system as its official system of weights and measures and the newly formed Kingdom of Italy likewise, following the lead given by Piedmont, adopted the metric system in 1861.

The *Exposition Universelle (1867)* (Paris exhibition) devoted a stand to the metric system and by 1875 two thirds of the European population and close on half the world's population had adopted the metric system. The principal European countries not to have adopted the metric system were Russia and the United Kingdom.

International Standards

In 1861, a committee of the British Association for Advancement of Science (BAAS) including William Thomson (later Lord Kelvin), James Clerk Maxwell and James Prescott Joule introduced the concept of a coherent system of units based on the metre, gram and second which, in 1873, was extended to include electrical units.

On 20 May 1875 an international treaty known as the *Convention du Mètre* (Metre Convention) was signed by 17 states. This treaty established the following organisations to conduct international activities relating to a uniform system for measurements:

- Conférence générale des poids et mesures (CGPM), an intergovernmental conference of official delegates of member nations and the supreme authority for all actions;
- Comité international des poids et mesures (CIPM), consisting of selected scientists and metrologists, which prepares and executes the decisions of the CGPM and is responsible for the supervision of the International Bureau of Weights and Measures;
- Bureau international des poids et mesures (BIPM), a permanent laboratory and world centre of scientific metrology, the activities of which include the establishment of the basic standards and

scales of the principal physical quantities and maintenance of the international prototype standards.

In 1881, first International Electrical Congress adopted the BAAS recommendations on electrical units, followed by a series of congresses in which further units of measure were defined.

Variants

A number of variants of the metric system evolved, all using the *Mètre des Archives* and *Kilogramme des Archives* as their base units, but differing in the definitions of the various derived units.

Centimetre-gram-second Systems

The centimetre gram second system of units (CGS) was the first coherent metric system, having been developed in the 1860s and promoted by Maxwell and Thomson. In 1874 this system was formally promoted by the British Association for the Advancement of Science (BAAS). The system's characteristics are that density is expressed in g/cm^3, force expressed in dynes and mechanical energy inergs. Thermal energy was defined in calories, one calorie being the energy required to raise the temperature of one gram of water from 15.5 °C to 16.5 °C. The meeting also proposed two sets of units for electrical and magnetic properties - the electrostatic set of units and the electromagnetic set of units.

Metre-kilogram-second Systems

The cgs units of electricity were cumbersome to work with. This was remedied at the 1893 International Electrical Congress held in Chicago by defining the "international" ampere and ohm using definitions based on the metre, kilogram and second. In 1901, Giorgi showed that by adding an electrical unit as a fourth base unit, the various anomalies in electromagnetic systems could be resolved. The metre-kilogram-second-coulomb (MKSC) and metre-kilogram-second-ampere (MKSA) systems are examples of such systems. The International System of Units (*System international units* or SI) is the current international standard metric system and the system most widely used around the world. It is an extension of Giorgi's MKSA system; its base units are the metre, kilogram, second, ampere, kelvin, candela and mole. Proposals have been made to change the definitions of four of the base units at the 24th meeting of the CGPM in October 2011. These changes should not affect the average person.

Metre-tonne-second Systems

The metre-tonne-second system of units (MTS) was based on the metre, tonne and second - the unit of force is the sthène and the unit of pressure is the pièze. It was invented in France in industry and was mostly used in the Soviet Union from 1933 to 1955.

Gravitational Systems

Gravitational metric systems use the kilogram-force (kilopond) as a base unit of force, with mass measured in a unit known as the hyl, TME, mug or metric slug. Note these are not part of the International System of Units (SI).

International System of Units

The 9th CGPM met in 1948, fifteen years after the 8th CGPM. In response to formal requests made by the International Union of Pure and Applied Physics and by the French Government to establish a practical system of units of measure, the CGPM requested the CIPM to prepare recommendations for a single practical system of units of measurement, suitable for adoption by all countries adhering to the Metre Convention. At the same time the CGPM formally adopted a recommendation for the writing and printing of unit symbols and of numbers. The recommendation also catalogued the recommended symbols for the most important MKS and CGS units of measure and for the first time the CGPM made recommendations concerning derived units.

The CIPM's draft proposal, which was an extensive revision and simplification of the metric unit definitions, symbols and terminology based on the MKS system of units, was put to the 10th CGPM in 1954. In accordance with Giorgi's proposals of 1901, the CIPM also recommended that the ampere be the base unit from which electromechanical would be derived. The definitions for the ohm and volt that had previously been in use were discarded and these units became derived units based on the metre, ampere, second and kilogram. After negotiations with the CIS and IUPAP, two further base units, the degree kelvin and the candela were also proposed as base units. The full system and name "Système International d'Unités" were adopted at the 11th CGPM. During the years that followed the definitions of the base units and particularly the *mise en pratique* to realise these definitions have been refined. The formal definition of International System of Units (SI) along with the associated resolutions passed by the CGPM and the CIPM are published by the BIPM on the Internet and in brochure form at regular intervals. The eighth edition of the brochure *Le Système International d'Unités - The International System of Units* was published in 2006.

"New SI"

When the metre was redefined in 1960, the kilogram was the only SI base unit that relied on a specific artifact. Moreover, after the 1996-1998 recalibration a clear divergence between the various prototype kilograms was observed.

At its 23rd meeting (2007), the CGPM mandated the CIPM to investigate the use of natural constants as the basis for all units of measure rather than the artifacts that were then in use. At a meeting of the CCU held in Reading, United Kingdom in September 2010, a resolution and draft changes to the SI brochure that were to be presented to the next meeting of the CIPM in October 2010 were agreed to in principle. The proposals that the CCU put forward were:

- In addition to the speed of light, four constants of nature - Planck's constant, an elementary charge, Boltzmann constant and Avogadro's number be defined to have exact values.
- The international prototype kilogram be retired
- The current definitions of the kilogram, ampere, kelvin and mole be revised.
- The wording of the definitions of all the base units be tightened up.

The CIPM meeting of October 2010 found that "the conditions set by the General Conference at its 23rd meeting have not yet been fully met. For this reason the CIPM does not propose a revision of the SI at the present time". The CIPM did however sponsor a resolution at the 24th CGPM in which the changes were agreed in principal and which were expected to be finalised at the CGPM's next meeting in 2014.

Usage Around the World

The usage of the metric system varies around the world. According to the American Central Intelligence Agency's *Factbook*, the International System of Units is the official system of measurement for all nations in the world except for Burma, Liberia and the United States. Some sources, though, suggest that this information is out of date: an Agence France-Presse of 2010 noted that Sierra Leone was to adopt the metric system, thereby aligning her system of measurement with her Mano River Union (MRU) neighbours Guinea and Liberia. Reports from Burma suggest that country is planning to adopt the metric system.

In the United States, where the use of metric units was by authorised by Congress in 1866, such units are widely used in science, military, and partially in industry, but customary units predominate in household use. At retail stores, the litre is a commonly used unit for volume, especially on bottles of beverages, and milligrams are used to denominate the amounts of medications, rather than grains. Also, other standardised measuring systems other than metric are still in universal international use, such as nautical miles and knots in international aviation.

In the countries of the Commonwealth of Nations the metric system has replaced the imperial system by varing degrees: Australia, New Zealand and Commonwealth countries in Africa are almost totally metric, India is mostly metric, Canada is partly metric while in the United Kingdom the metric system, the use of which was first permitted for trade in 1864, is used in government work, in most professional applications including building, health and engineering and for pricing in most application, both wholesale and retail. However the imperial system continues to be used in many unregulated and casual everyday applications.

A number of other jurisdictions, such as Hong Kong, have laws mandating or permitting other systems of measurement in parallel with the metric system in some or all contexts.

Variations in Spelling

Although the symbol for the kilometre throughout the world is "km", there is no consistency in the spelling of the name "kilometre". Similar variations are found with the spelling of other units of measure in various countries including differences in American English and British spelling. For example *metre* and *litre* are used in the United States whereas *metre* and *litre* are used in other English-speaking countries. In addition, the official US spelling for the rarely used SI prefix for ten is *deka*. In American English the term *metric ton* is the normal usage whereas in other varieties of English *tonne* is common. *Gram* is also sometimes spelled *gramme* in English-speaking countries other than the United States, though this older usage is declining. The US government has approved this terminology for official use. In scientific contexts, only the symbols are used; since these are universally the same, the differences do not arise in practice in scientific use.

Conversion and Calculation Errors

The dual usage of metric and non metric units has resulted in serious errors. These include:

- The degree of overloading of an American International Airways aircraft flying from Miami to Maiquetia, Venezuela on 26 May 1994 was consistent with a cargo weighed in kilograms, not pounds.
- The Institute for Safe Medication Practices has reported that confusion between grains and grams led to a patient receiving phenobarbital 0.5 grams instead of 0.5 grains (0.03 grams) after the practitioner misread the prescription.
- The Canadian "Gimli Glider" accident in 1983, when a Boeing 767 jet ran out of fuel in mid-flight because of two mistakes in calculating the fuel supply of Air Canada's first aircraft to use metric measurements.
- The root cause of NASA's 1999 loss of the $125 million Mars Climate Orbiter which crashed into Mars was a mismatch of units - the spacecraft engineers calculated the thrust forces required for velocity changes using US customary units (lbf·s) whereas the team who built the thrusters were expecting a value in metric units (N·s) as per the agreed specification.

8

Unit of Measurement

A unit of measurement is a definite magnitude of a physical quantity, defined and adopted by convention and/or by law, that is used as a standard for measurement of the same physical quantity. Any other value of the physical quantity can be expressed as a simple multiple of the unit of measurement. For example, length is a physical quantity. The metre is a unit of length that represents a definite predetermined length. When we say 10 metres (or 10 m), we actually mean 10 times the definite predetermined length called "metre".

The definition, agreement, and practical use of units of measurement have played a crucial role in human endeavour from early ages up to this day. Different systems of units used to be very common. Now there is a global standard, the International System of Units (SI), the modern form of the metric system.

In trade, weights and measures is often a subject of governmental regulation, to ensure fairness and transparency. The Bureau international des poids et measures (BIPM) is tasked with ensuring worldwide uniformity of measurements and their traceability to the International System of Units (SI). Metrology is the science for developing nationally and internationally accepted units of weights and measures.

In physics and metrology, units are standards for measurement of physical quantities that need clear definitions to be useful. Reproducibility of experimental results is central to the scientific method. A standard system of units facilitates this. Scientific systems of units are a refinement of the concept of weights and measures developed long ago for commercial purposes.

Science, medicine, and engineering often use larger and smaller units of measurement than those used in everyday life and indicate them more precisely. The judicious selection of the units of measurement can aid researchers in problem solving. In the social sciences, there are no standard units of measurement and the theory and practice of measurement is studied in psychometrics and the theory of conjoint measurement.

A unit of measurement is a standardised quantity of a physical property, used as a factor to express occurring quantities of that property. Units of measurement were among the earliest tools invented by humans. Primitive societies needed rudimentary measures for many tasks: constructing dwellings of an appropriate size and shape, fashioning clothing, or bartering food or raw materials.

The earliest known uniform systems of weights and measures seem to have all been created sometime in the 4th and 3rd millennia BC among the ancient peoples of Mesopotamia, Egypt and the Indus Valley, and perhaps also Elam inPersia as well.

In "The Magna Carta" of 1215 (The Great Charter) with the seal of King John, put before him by the Barons of England, King John agreed in Clause 35 "There shall be one measure of wine throughout our whole realm, and one measure of ale and one measure of corn—namely, the London quart;—and one width of dyed and russet and hauberk cloths—namely, two ells below the selvage....". The Magna Carta helped lay the foundations of freedom codified in English Law and subsequently American Law. Many systems were based on the use of parts of the body and the natural surroundings as measuring instruments. Our present knowledge of early weights and measures comes from many sources.

Systems of Units

Traditional Systems

Prior to the near global adoption of the metric system many different systems of measurement had been in use. Many of these were related to some extent or other. Often they were based on the dimensions of the human body according to the proportions described by Marcus Vitruvius Pollio. As a result, units of measure could vary not only from location to location, but from person to person.

Metric Systems

A number of metric systems of units have evolved since the adoption of the original metric system in France in 1791. The current international standard metric system is the International System of Units. An important feature of modern systems is standardisation. Each unit has a universally recognised size.

Both the Imperial units and US customary units derive from earlier English units. Imperial units were mostly used in the British Commonwealth and the former British Empire. US customary units are still the main system of measurement used in the United States despite Congress having legally authorised metric measure on 28 July 1866. Some steps towards US metrication have been made, particularly the redefinition of basic US units to derive exactly from SI units, so that in the US the inch is now defined as 0.0254 m (exactly), and the avoirdupois pound is now defined as 453.59237 g (exactly)

Natural Systems

While the above systems of units are based on arbitrary unit values, formalised as standards, some unit values occur naturally in science. Systems of units based on these are called natural units. Similar to natural units, atomic units (au) are a convenient system of units of measurement used in atomic physics.

Also a great number of unusual and non-standard units may be encountered. These may include the Solar mass, the Megaton (1,000,000 tons of TNT).

Legal Control of Weights and Measures

To reduce the incidence of retail fraud, many national statutes have standard definitions of weights and measures that may be used (hence "statute measure"), and these are verified by legal officers.

Base and Derived Units

Different systems of units are based on different choices of a set of fundamental units. The most widely used system of units is the International System of Units, or SI. There are seven SI base units. All other SI units can be derived from these base units.

For most quantities a unit is absolutely necessary to communicate values of that physical quantity. For example, conveying to someone a particular length without using some sort of unit is impossible, because a length cannot be described without a reference used to make sense of the value given.

But not all quantities require a unit of their own. Using physical laws, units of quantities can be expressed as combinations of units of other quantities. Thus only a small set of units is required. These units are taken as the *base units*. Other units are *derived units*. Derived units are a matter of convenience, as they can be expressed in terms of basic units. Which units are considered base units is a matter of choice. The base units of SI are actually not the smallest set possible. Smaller sets have been defined. For example, there are unit sets in which the electric and magnetic field have the same unit. This is based on physical laws that show that electric and magnetic field are actually different manifestations of the same phenomenon.

Calculations with Units

Units as Dimensions

Any value of a physical quantity is expressed as a comparison to a unit of that quantity. For example, the value of a physical quantity Z is expressed as the product of a unit [Z] and a numerical factor:

$Z = n \times [Z] = n[Z]$. For example, "2 candlesticks" Z = 2 [candlestick].

The multiplication sign is usually left out, just as it is left out between variables in scientific notation of formulas. The conventions used to express quantities is referred to as quantity calculus. In formulas the unit [Z] can be treated as if it were a specific magnitude of a kind of physical dimension.

Units can only be added or subtracted if they are the same type; however units can always be multiplied or divided, as George Gamow used to explain:

"2 candlesticks" times "3 cabdrivers" = 6 [candlestick][cabdriver].

A distinction should be made between units and standards. A unit is fixed by its definition, and is independent of physical conditions such as temperature. By contrast, a standard is a physical realisation of a unit, and realises that unit only under certain physical conditions. For example, the metre is a unit, while a metal bar is a standard. One metre is the same length regardless of temperature, but a metal bar will be one metre long only at a certain temperature.

Guidelines

- Treat units algebraically. Only add like terms. When a unit is divided by itself, the division yields a unitless one. When two different units are multiplied, the result is a new unit, referred to by the combination of the units. For instance, in SI, the unit of speed is metres per second (m/s). A unit can be multiplied by itself, creating a unit with an exponent (e.g. m^2/s^2). Put simply, units obey the laws of indices.

- Some units have special names, however these should be treated like their equivalents. For example, one newton (N) is equivalent to one kg·m/s². Thus a quantity may have several unit designations, for example: the unit for surface tension can be referred to as either N/m (newtons per metre) or kg/s² (kilograms per second squared). Whether these designations are equivalent is disputed amongst metrologists.

Expressing a Physical Value in Terms of Another Unit

Conversion of units involves comparison of different standard physical values, either of a single physical quantity or of a physical quantity and a combination of other physical quantities.

Starting with:

$$Z = n_i \times [Z]_i$$

just replace the original unit $[Z]_i$ with its meaning in terms of the desired unit $[Z]_j$, e.g. if $[Z]_i = c_{ij} \times [Z]_j$ then:

$$Z = n_i \times (c_{ij} \times [Z]_j) = (n_i \times c_{ij}) \times [Z]_j$$

Now n_i and c_{ij} are both numerical values, so just calculate their product.

Or, which is just mathematically the same thing, multiply Z by unity, the product is still Z:

$$Z = n_i \times [Z]_i \times (c_{ij} \times [Z]_j / [Z]_i)$$

For example, you have an expression for a physical value Z involving the unit *feet per second* ($[Z]_i$) and you want it in terms of the unit *miles per hour* ($[Z]_j$):

1. Find facts relating the original unit to the desired unit:

   1 mile = 5280 feet and 1 hour = 3600 seconds

2. Next use the above equations to construct a fraction that has a value of unity and that contains units such that, when it is multiplied with the original physical value, will cancel the original units:

$$1 = \frac{1\text{mi}}{5280\text{ft}} \quad \text{and} \quad 1 = \frac{3600\text{s}}{1\text{h}}$$

3. Last, multiply the original expression of the physical value by the fraction, called a *conversion factor*, to obtain the same physical value expressed in terms of a different unit. Note: since valid conversion factors are dimensionless and have a

numerical value of one, multiplying any physical quantity by such a conversion factor (which is 1) does not change that physical quantity.

$$52.8\frac{\text{ft}}{\text{s}} = 52.8\frac{\text{ft}}{\text{s}}\frac{1\text{mi}}{5280\text{ft}}\frac{3600\text{s}}{1\text{h}} = \frac{52.8\times3600}{5280}\text{mi/h} = 36\text{mi/h}$$

Or as an example using the metric system, you have a value of fuel economy in the unit *litres per 100 kilometres* and you want it in terms of the unit *microlitres per metre*:

$$\frac{9\text{L}}{100\text{km}} = \frac{9\text{L}}{100\text{km}}\frac{1000000\mu\text{L}}{1\text{L}}\frac{1\text{km}}{1000\text{m}} = \frac{9\times1000000}{100\times1000}\mu\text{L/m} = 90\mu\text{L/m}$$

Real-world Implications

One example of the importance of agreed units is the failure of the NASA Mars Climate Orbiter, which was accidentally destroyed on a mission to the planet Mars in September 1999 instead of entering orbit, due to miscommunications about the value of forces: different computer programs used different units of measurement (newton versus pound force). Enormous amounts of effort, time, and money were wasted. On April 15, 1999 Korean Air cargo flight 6316 from Shanghai to Seoul was lost due to the crew confusing tower instructions (in metres) and altimetre readings (in feet). Three crew and five people on the ground were killed. Thirty seven were injured.

In 1983, a Boeing 767 (which came to be known as the Gimli Glider) ran out of fuel in mid-flight because of two mistakes in figuring the fuel supply of Air Canada's first aircraft to use metric measurements. This accident is apparently the result of confusion both due to the simultaneous use of metric & Imperial measures as well as mass & volume measures.

Patent Drawing

A patent application or patent may contain drawings, also called patent drawings, illustrating the invention, some of its embodiments (which are particular implementations or methods of carrying out the invention), or the prior art. The drawings may be required by the law to be in a particular form, and the requirements may vary depending on the jurisdiction.

Jurisdictions

Europe: Under the European Patent Convention, Article 78(1) EPC provides that a European patent application shall contain any drawings referred to in the description or the claims. Drawings are therefore optional. Rule 46 EPC specifies the form in which the drawings must be executed.

The European search report is drawn up in respect of a European patent application on the basis of the claims, with due regard to the description and any drawings. In addition, the extent of the protection conferred by a European patent or a European patent application is determined by the claims, with the description and drawings being used to interpret the claims.

Patent Cooperation Treaty

Under the Patent Cooperation Treaty, Article 7 PCT notably provides that the drawings are required when they are necessary for the understanding of the invention. Rule 11.13 PCT specifies special physical requirements for drawings in an international application.

United States

In the United States, the applicant for a patent is required by law to furnish a drawing of the invention whenever the nature of the case requires a drawing to understand the invention. This drawing must be filed with the application. This includes practically all inventions except compositions of matter or processes, but a drawing may also be useful in the case of many processes.

The drawing must show every feature of the invention specified in the claims, and is required by the U.S. patent office rules to be in a particular form. The United States Patent and Trademark Office (USPTO) specifies the size of the sheet on which the drawing is made, the type of paper, the margins, and other details relating to the making of the drawing. The reason for specifying the standards in detail is that the drawings are printed and published in a uniform style when the patent issues, and the drawings must also be such that they can be readily understood by persons using the patent descriptions.

No names or other identification are permitted within the "sight" of the drawing, and applicants are expected to use the space above and between the hole locations to identify each sheet of drawings. This identification may consist of the attorney's name and docket number or the inventor's name and application number and may include the sheet number and the total number of sheets filed (for example, "sheet 2 of 4"). The following rule, reproduced from title 37 of the Code of Federal Regulations, relates to the standards for drawings:

History

From 1790 to 1880 in the US, patent models were required. A patent model was a scratch-built miniature model no larger than 12" by 12" by 12", approximately 30 cm by 30 cm by 30 cm, that showed how an invention works. Some inventors still willingly submitted models at the turn of the twentieth century. In some cases, an inventor may still want to present a "working model" as an evidence to prove actual reduction to practice in aninterference proceeding. In some jurisdictions patent models stayed an aid to demonstrate the operation of the invention. In applications involving genetics, samples of genetic material or DNA sequences may be required.

Figure: *Patent model for loom.*

The United States patent law was revised in 1793. It stated that the Commissioner of the USPTO could ask for additional information, drawings, or diagrams if the description is not clear. By then, the rate of patent grants had grown to about 20 per year and the time burden on the Secretary of State was considered to be too burdensome. Patent applications were no longer examined. Patents were granted simply by submitting a written description of an invention, a model of the invention, if appropriate, and paying a fee of $30 then, and now $1000 in 2006 US dollars.

Drawings and Photographs

In utility and design patent applications, drawings can be in black ink or colour. Black and white drawings are normally required. On rare occasions, colour drawings may be necessary as the only practical medium by which to disclose the subject matter sought to be patented in a utility or design patent application or the subject matter of a statutory invention registration.

Black and white photographs are not ordinarily permitted in utility and design patent applications, unless this is the only practicable medium for illustrating the claimed invention. For example, photographs of electrophoresis gels, blots, autoradiographs, cell cultures, histological tissue cross sections, animals, plants, in vivo imaging, etc. Colour photographs can be accepted in utility and design patent applications if the conditions for accepting colour drawings and black and white photographs have been satisfied.

Features

Patent drawing features can contain the following features:

- Identification of drawings: includes the title of the invention, inventor's name, and application number.. etc.
- Graphic forms in drawings. Chemical or mathematical formulae, tables, and waveforms may be submitted as drawings and are subject to the same requirements as drawings. Each chemical or mathematical formula must be labelled as a separate figure, using brackets when necessary, to show that information is properly integrated.
- Type of paper: generally flexible, strong, white, smooth, non-shiny, and durable.
- Size of paper: Must be the same size DIN size A4, (8 1/2 by 11 inches).
- Some kind of Margin standard.
- Views. The drawing must contain as many views as necessary to show the invention. The views may be plan, elevation, section, or perspective views.
- Arrangement of views: All views on the same sheet in the same direction.
- Front page view
- Scale: large enough to show the mechanism
- Shading: aids in understanding the invention used to indicate the surface or shape of spherical, cylindrical, and conical elements of an object.
- Symbols: Graphical drawing symbols may be used for conventional elements when appropriate.
- Legends: should contain as few words as possible.
- Numbers, letters, and reference characters.
- Lead lines: between the reference characters and the details referred to, and

- Arrows: at the ends of lines, provided that their meaning is cleared.
- International drawing formats: Format requirements differ by country where the patent is being filed. Check local patent offices for format requirements.

The patent drawing can further contain a numbering of sheets of drawings, numbering of views, copyright notice, security markings, corrections (durable and permanent), no holes, and a type of drawing indication.

Views

The views in the drawing may be plan, elevation, section, or perspective views:

- Exploded views: views with the separated parts embraced by a bracket, to show the relationship or order of assembly of various parts are permissible.
- Partial views: a view of a large machine or device in its entirety may be broken into partial views on a single sheet, or extended over several sheets if there is no loss in facility of understanding the view.
- Sectional views: The plane upon which a sectional view is taken should be indicated on the view from which the section is cut by a broken line.
- Alternate position: A moved position may be shown by a broken line superimposed upon a suitable view if this can be done without crowding; otherwise, a separate view must be used for this purpose, and
- Modified forms. Modified forms of construction must be shown in separate views.

Specification (Technical Standard)

A specification (often abbreviated as spec) is an explicit set of requirements to be satisfied by a material, product, or service. Should a material, product or service fail to meet one or more of the applicable specifications, it may be referred to as being *out of specification*; the abbreviation OOS may also be used. Specs are a type of technical standard.

A technical specification may be developed by any of various kinds of organisations, both public and private. Example organisation types include a corporation, a consortium (a small group of corporations), a trade association (an industry-wide group of corporations), a national government (including its military, regulatory agencies, and national laboratories and institutes), a professional association (society), or a purpose-made standards organisation such as ISO. It is common for one organisation to *refer to* (*reference, call out, cite*) the standards of another. Voluntary standards may become mandatory if adopted by a government or business contract.

Sometimes the term *specification* is used in connection with a data sheet (or *spec sheet*). A data sheet describes the technical characteristics of an item or product. It can be published by a manufacturer to help people choose products or to help use the products.

In engineering, manufacturing, and business, it is vital for suppliers, purchasers, and users of materials, products, or services to understand and agree upon all requirements. A specification is a type of a standard which is often referenced by a contract or procurement document. It provides the necessary details about the specific requirements.

Specifications may be written by government agencies, standards organisations (ASTM, ISO, CEN, DoD, etc.), trade associations, corporations, and others.

A product specification does not necessarily prove a product to be correct. An item might be verified to comply with a specification or stamped with a specification number: This does not, by itself, indicate that the item is fit for any particular use. The people who use the item (engineers, trade unions, etc.) or specify the item (building codes, government, industry, etc.) have the responsibility to consider the choice of available specifications, specify the correct one, enforce compliance, and use the item correctly. Validation of suitability is necessary.

Guidance and Content

Sometimes a guide or a standing operating procedure is available to help write and format a good specification. A specification might include:

- Descriptive title, number, identifier, etc. of the specification
- Date of last effective revision and revision designation
- A logo or trademark to indicate the document copyright, ownership and origin
- Table of Contents (TOC), if the document is long

- Person, office, or agency responsible for questions on the specification, updates, and deviations.
- The significance, scope or importance of the specification and its intended use.
- Terminology, definitions and abbreviations to clarify the meanings of the specification
- Test methods for measuring all specified characteristics
- Material requirements: physical, mechanical, electrical, chemical, etc. Targets and tolerances.
- Acceptance testing, including Performance testing requirements. Targets and tolerances.
- Drawings, photographs, or technical illustrations
- Workmanship
- Certifications required.
- Safety considerations and requirements
- Environmental considerations and requirements
- Quality control requirements, acceptance sampling, inspections, acceptance criteria
- Person, office, or agency responsible for enforcement of the specification.
- Completion and delivery.
- Provisions for rejection, reinspection, rehearing, corrective measures
- References and citations for which any instructions in the content maybe required to fulfill the traceability and clarity of the document
- Signatures of approval, if necessary
- Change record to summarize the chronological development, revision and completion if the document is to be circulated internally
- Annexes and Appendices that are expand details, add clarification, or offer options.

Process Capability Considerations

A good engineering specification, by itself, does not necessarily imply that all products sold to that specification actually meet the listed targets and tolerances. Actual production of any material, product, or service involves inherent variation of output. With a normal distribution, the tails of production may extend well beyond plus and minus three standard deviations from the process average.

The process capability of materials and products needs to be compatible with the specified engineering tolerances. Process controls must be in place and an effective Quality management system, such as Total Quality Management, needs to keep actual production within the desired tolerances.

Effective enforcement of a specification is necessary for it to be useful.

Construction Specifications in North America

Specifications in North America form part of the contract documents that accompany and govern the construction of a building. The guiding master document is the latest edition of MasterFormat. It is a consensus document that is jointly sponsored by two professional organisations: Construction Specifications Canada and Construction Specifications Institute.

While there is a tendency to believe that "Specs overrule Drawings" in the event of discrepancies between the text document and the drawings, the actual intent—made explicit in the contract between the Owner and the Contractor—is for the drawings and specifications to be complementary, together providing the information required for a complete facility.

The Specifications fall into 50 Divisions, or broad categories of work results involved in construction. The Divisions are subdivided into Sections, each one addressing a narrow scope of the construction work. For instance, firestopping is addressed in Section 078400 - Firestopping. It forms part of Division 07, which is Thermal and Moisture Protection. Division 07 also addresses building envelope and fireproofing work.

Each Section is subdivided into three distinct Parts: "General", "Products" and "Execution". The MasterFormat system can be successfully applied to residential, commercial, civil, and industrial construction.

Specifications can be either "performance-based", whereby the specifier restricts the text to stating the performance that must be achieved by the completed work, or "prescriptive", whereby the specifier indicates specific products, vendors and even contractors that are acceptable for each workscope. Most construction specifications are a combination of performance-based and prescriptive types, naming acceptable manufacturers and products while also specifying certain standards and design criteria that must be met.

While North American specifications are usually restricted to broad descriptions of the work, European ones can include actual work quantities, including such things as area of drywall to be built in square metres, like a bill of materials. This type of specification is a collaborative effort between a specwriter and a quantity surveyor. This approach is unusual in North America, where each bidder performs a quantity survey on the basis of both drawings and specifications.

Although specifications are usually issued by the architect's office, specwriting itself is undertaken by the architect and the various engineers or by specialist specwriters. Specwriting is often a distinct professional trade, with professional certifications such as "Certified Construction Specifier" (CCS) through the professional organisations noted above. Specwriters are either employees of or sub-contractors to architects, engineers, or construction management companies. Specwriters frequently meet with manufacturers of building materials who seek to have their products specified on upcoming construction projects so that contractors can include their products in the estimates leading to their proposals.

Construction Specifications in the UK

Specifications in the UK are prepared by construction professionals such as Architects, Structural Engineers, Landscape Architects and Building Services Engineers. They are created from previous project specifications, in-house documents or master specifications such as the National Building Specification (NBS). The National Building Specification is owned by the Royal Institute of British Architects (RIBA) through their commercial group RIBA Enterprises (RIBAe). NBS master specifications provide content that is broad and comprehensive, and delivered using software functionality that enables specifiers to customize the content to suit the needs of the project and to keep up to date. UK project specification types fall into two main categories prescriptive and performance. Prescriptive specifications define the requirements using generic or proprietary descriptions of what is required, whereas as performance specifications focus on the outcomes rather than the characteristics of the components. Specifications are an integral part of Building Information Modelling and cover the non-geometric requirements.

Food and Drug Specifications

Pharmaceutical products can usually be tested and qualified by various Pharmacopoeia. Current existing pronounced standards include:

- British Pharmacopoeia
- European Pharmacopoeia
- Japanese Pharmacopoeia
- The International Pharmacopoeia
- United States Pharmacopoeia.

If any pharmaceutical product is not covered by the above standards, it can be evaluated by the additional source of Pharmacopoeia from other nations, from industrial specifications. or from standardised formulary such as;

- British National Formulary for Children
- British National Formulary
- National Formulary.

A similar approach is adopted by the food manufacturing, of which Codex Alimentarius ranks the hightest standards, followed by regional and national standards.

The coverage of food and drug standards by ISO is currently less fruitful and not yet put forward as an urgent agenda due to the tight restrictions of regional or national constitution

Specifications and other standards can be externally imposed as discussed above, but also intenal manufacturing and quality specifications. These exist not only for the food or pharmaceutical product but also for the processing machinery, quality processes, packaging, logistics (cold chain), etc. and are examplified by *ISO 14134* and *ISO 15609*.

The converse of explicit statement of specifications is a process for dealing with observations that are out-of-specification. The United States Food and Drug Administration has published a non-binding recommendation that addresses just this point.

At the present time, much of the information and regulations concerning food and food products remain in a form which makes it difficult to apply automated information processing, storage and transmission methods and techniques.

Data systems that can process, store and transfer information about food and food products need formal specifications for the representations of data about food and food products in order to operate effectively and efficiently.

Development of formal specifications for food and drug data with the necessary and sufficient clarity and precision for use specifically by digital computing systems have begun to emerge from government agencies and standards organisations.

The United States Food and Drug Administration has published specifications for a "Structured Product Label" which drug manufacturers must by mandate use to submit electronically the information on a drug label.

Recently, ISO has made some progress in the area of food and drug standards and formal specifications for data about regulated substances through the publication of *ISO 11238*

Information Technology

Formal Specification

A formal specification is a mathematical description of software or hardware that may be used to develop an implementation. It describes *what* the system should do, not (necessarily) *how* the system should do it. Given such a specification, it is possible to use formal verification techniques to demonstrate that a candidate system design is correct with respect to the specification. This has the advantage that incorrect candidate system designs can be revised before a major investment has been made in actually implementing the design. An alternative approach is to use provably correct refinement steps to transform a specification into a design, and ultimately into an actual implementation, that is correct by construction.

Program Specification

A program specification is the definition of what a computer program is expected to do. It can be *informal*, in which case it can be considered as a blueprint or user manual from a developer point of view, or *formal*, in which case it has a definite meaning defined in mathematical or programmatic terms. In practice, many successful specifications are written to understand and fine-tune applications that were already well-developed, although safety-critical software systems are often carefully specified prior to application development. Specifications are most important for external interfaces that must remain stable.

Functional Specification

In software development, a functional specification (also, functional spec or specs or functional specifications document (FSD)) is the set of documentation that describes the behaviour of a computer program or larger software system. The documentation typically describes various inputs that can be provided to the software system and how the system responds to those inputs.

Web Service Specification

Web services specifications are often under the umbrella of a quality management system.

Document Specification

These types of documents define how a specific document should be written, which may include, but is not limited to, the systems of a document naming, version, layout, referencing, structuring, appearance, language, copyright, hierarchy or format, etc. Very often, this kind of specifications is complemented by a designated template.

Infrastructure

Infrastructure is basic physical and organisational structures needed for the operation of a society or enterprise, or the services and facilities necessary for an economy to function. It can be generally defined as the set of interconnected structural elements that provide framework supporting an entire structure of development.

The term typically refers to the technical structures that support a society, such as roads, water supply, sewers, electrical grids, telecommunications, and so forth, and can be defined as "the physical components of interrelated systems providing commodities and services essential to enable, sustain, or enhance societal living conditions."

Viewed functionally, infrastructure *facilitates* the production of goods and services, and also the distribution of finished products to markets, as well as basic social services such as schools and hospitals; for example, roads enable the transport of raw materials to a factory. In military parlance, the term refers to the buildings and permanent installations necessary for the support, redeployment, and operation of military forces.

According to the *Online Etymology Dictionary*, the word infrastructure has been used in English since at least 1927, originally meaning "The installations that form the basis for any operation or system".

Other sources, such as the *Oxford English Dictionary*, trace the word's origins to earlier usage, originally applied in a military sense. The word was imported from French, where it means *subgrade*, the native material underneath a constructed pavement or railway. The word is a combination of the Latin prefix "infra", meaning "below", and "structure". The military use of the term achieved currency in the United States after the formation of NATO in the 1940s, and was then adopted by urban planners in its modern civilian sense by 1970.

The term came to prominence in the United States in the 1980s following the publication of *America in Ruins*, which initiated a public-policy discussion of the nation's "infrastructure crisis", purported to be caused by decades of inadequate investment and poor maintenance of public works. This crisis discussion as contributed to the increase in infrastructure asset management and maintenance planning in the US.

That public-policy discussion was hampered by lack of a precise definition for infrastructure. A US National Research Council panel sought to clarify the situation by adopting the term "public works infrastructure", referring to:

> *"... both specific functional modes – highways, streets, roads, and bridges; mass transit; airports and airways; water supply and water resources; wastewater management; solid-waste treatment and disposal; electric power generation and transmission; telecommunications; and hazardous waste management – and the combined system these modal elements comprise. A comprehension of infrastructure spans not only these public works facilities, but also the operating procedures, management practices, and development policies that interact together with societal demand and the physical world to facilitate the transport of people and goods, provision of water for drinking and a variety of other uses, safe disposal of society's waste products, provision of energy where it is needed, and transmission of information within and between communities."*

In Keynesian economics, the word *infrastructure* was exclusively used to describe public assets that facilitate production, but not private assets of the same purpose. In post-Keynesian times, however, the word has grown in popularity. It has been applied with increasing generality to suggest the internal framework discernible in any technology system or business organisation.

"Hard" versus "Soft" Infrastructure

In this article, "hard" infrastructure refers to the large physical networks necessary for the functioning of a modern industrial nation, whereas "soft" infrastructure refers to all the institutions which are required to maintain the economic, health, and cultural and social standards of a country, such as the financial system, the education system, the health care system, the system of government, and law enforcement, as well as emergency services.

Types of Hard Infrastructure

The following list of hard infrastructure is limited to capital assets that serve the function of conveyance or channelling of people, vehicles, fluids, energy, or information, and which take the form either of a network or of a critical node used by vehicles, or used for the transmission of electro-magnetic waves.

Infrastructure systems include both the fixed assets, and the control systems and software required to operate, manage and monitor the systems, as well as any accessory buildings, plants, or vehicles that are an essential part of the system. Also included are fleets of vehicles operating according to schedules such as public transit buses and garbage collection, as well as basic energy or communications facilities that are not usually part of a physical network, such as oil refineries, radio, and television broadcasting facilities.

Transportation Infrastructure

- Road and highway networks, including structures (bridges, tunnels, culverts, retaining walls), signage and markings, electrical systems (street lighting and traffic lights), edge treatments (curbs, sidewalks, landscaping), and specialised facilities such as road maintenance depots and rest areas
- Mass transit systems (Commuter rail systems, subways, tramways, trolleys and bus transportation)
- Railways, including structures, terminal facilities (rail yards, train stations), level crossings, signalling and communications systems
- Canals and navigable waterways requiring continuous maintenance (dredging, etc)
- Seaports and lighthouses
- Airports, including air navigational systems
- Bicycle paths and pedestrian walkways
- Ferries.

For canals, railroads, highways, airways and pipelines see Grübler (1990), which provides a detailed discussion of the history and importance of these major infrastructures.

Energy Infrastructure

- Electrical power network, including generation plants, electrical grid, substations, and local distribution.
- Natural gas pipelines, storage and distribution terminals, as well as the local distribution network. Some definitions may include the gas wells, as well as the fleets of ships and trucks transporting liquefied gas.
- Petroleum pipelines, including associated storage and distribution terminals. Some definitions may include the oil wells, refineries, as well as the fleets of tanker ships and trucks.
- Specialised coal handling facilities for washing, storing, and transporting coal. Some definitions may include Coal mines.
- Steam or hot water production and distribution networks for district heating systems.
- Electric vehicle networks for charging electric vehicles.

Coal mines, oil wells and natural gas wells may be classified as being part of the mining and industrial sector of the economy, not part of infrastructure.

Water Management Infrastructure

- Drinking water supply, including the system of pipes, storage reservoirs, pumps, valves, filtration and treatment equipment and metres, including buildings and structures to house the equipment, used for the collection, treatment and distribution of drinking water
- Sewage collection, and disposal of waste water
- Drainage systems (storm sewers, ditches, etc)
- Major irrigation systems (reservoirs, irrigation canals)
- Major flood control systems (dikes, levees, major pumping stations and floodgates)
- Large-scale snow removal, including fleets of salt spreaders, snow-plows, snowblowers, dedicated dump-trucks, sidewalk plows, the dispatching and routing systems for these fleets, as well as fixed assets such as snow dumps, snow chutes, snow melters
- Coastal management, including structures such as seawalls, breakwaters, groynes, floodgates, as well as the use of soft engineering techniques such as beach nourishment, sand dune

stabilisation and the protection of mangrove forests and coastal wetlands.

Communications Infrastructure

- Postal service, including sorting facilities
- Telephone networks (land lines) including telephone exchange systems
- Mobile phone networks
- Television and radio transmission stations, including the regulations and standards governing broadcasting
- Cable television physical networks including receiving stations and cable distribution networks (does not include content providers or "networks" when used in the sense of a specialised channel such as CNN or MTV)
- The Internet, including the internet backbone, core routers and server farms, local internet service providers as well as the protocols and other basic software required for the system to function (does not include specific websites, although may include some widely-used web-based services, such as social network services and web search engines)
- Communications satellites
- Undersea cables
- Major private, government or dedicated telecommunications networks, such as those used for internal communication and monitoring by major infrastructure companies, by governments, by the military or by emergency services, as well as national research and education networks
- Pneumatic tube mail distribution networks.

Solid Waste Management

- Municipal garbage and recyclables collection
- Solid waste landfills
- Solid waste incinerators and plasma gasification facilities
- Materials recovery facilities
- Hazardous waste disposal facilities.

Earth Monitoring and Measurement Networks

- Meteorological monitoring networks
- Tidal monitoring networks

- Stream Gauge or fluviometric monitoring networks
- Seismometre networks
- Earth observation satellites
- Geodetic benchmarks
- Global Positioning System
- Spatial Data Infrastructure.

Types of Soft Infrastructure

Soft infrastructure includes both physical assets such as highly specialised buildings and equipment, as well as non-physical assets such as the body of rules and regulations governing the various systems, the financing of these systems, as well as the systems and organisations by which highly skilled and specialised professionals are trained, advance in their careers by acquiring experience, and are disciplined if required by professional associations (professional training, accreditation and discipline). Unlike hard infrastructure, the essence of soft infrastructure is the delivery of specialised services to people. Unlike much of the service sector of the economy, the delivery of those services depend on highly developed systems and large specialised facilities or institutions that share many of the characteristics of hard infrastructure.

Governance Infrastructure

- The system of government and law enforcement, including the political, legislative, law enforcement, justice and penal systems, as well as specialised facilities (government offices, courthouses, prisons, etc), and specialised systems for collecting, storing and disseminating data, laws and regulation
- Emergency services, such as police, fire protection, and ambulances, including specialised vehicles, buildings, communications and dispatching systems
- Military infrastructure, including military bases, arms depots, training facilities, command centres, communication facilities, major weapons systems, fortifications, specialised arms manufacturing, strategic reserves

Economic Infrastructure

- The financial system, including the banking system, financial institutions, the payment system, exchanges, the money supply, financial regulations, as well as accounting standards and regulations

- Major business logistics facilities and systems, including warehouses as well as warehousing and shipping management systems
- Manufacturing infrastructure, including industrial parks and special economic zones, mines and processing plants for basic materials used as inputs in industry, specialised energy, transportation and water infrastructure used by industry, plus the public safety, zoning and environmental laws and regulations that govern and limit industrial activity, and standards organisations
- Agricultural, forestry and fisheries infrastructure, including specialised food and livestock transportation and storage facilities, major feedlots, agricultural price support systems (including agricultural insurance), agricultural health standards, food inspection, experimental farms and agricultural research centres and schools, the system of licencing and quota management, enforcement systems against poaching, forest wardens, and fire fighting.

Social Infrastructure

- The health care system, including hospitals, the financing of health care, including health insurance, the systems for regulation and testing of medications and medical procedures, the system for training, inspection and professional discipline of doctors and other medical professionals, public health monitoring and regulations, as well as coordination of measures taken during public health emergencies such as epidemics
- The educational and research system, including elementary and secondary schools, universities, specialised colleges, research institutions, the systems for financing and accrediting educational institutions
- Social welfare systems, including both government support and private charity for the poor, for people in distress or victims of abuse.

Cultural, Sports and Recreational Infrastructure

- Sports and recreational infrastructure, such as parks, sports facilities, the system of sports leagues and associations
- Cultural infrastructure, such as concert halls, museums, libraries, theatres, studios, and specialised training facilities
- Business travel and tourism infrastructure, including both man-made and natural attractions, convention centres, hotels,

restaurants and other services that cater mainly to tourists and business travellers, as well as the systems for informing and attracting tourists, and travel insurance.

Engineering and Construction

Engineers generally limit the use of the term "infrastructure" to describe fixed assets that are in the form of a large network, in other words, "hard" infrastructure. Recent efforts to devise more generic definitions of infrastructure have typically referred to the network aspects of most of the structures, and to the accumulated value of investments in the networks as assets. One such effort defines infrastructure as the network of assets "where the system as a whole is intended to be maintained indefinitely at a specified standard of service by the continuing replacement and refurbishment of its components".

Civil Defence and Economic Development

Civil defence planners and developmental economists generally refer to both hard and soft infrastructure, including public services such as schools and hospitals, emergency services such as police and fire fighting, and basic financial services.

Military

Military strategists use the term infrastructure to refer to all building and permanent installations necessary for the support of military forces, whether they are stationed in bases, being deployed or engaged in operations, such as barracks, headquarters, airfields, communications facilities, stores of military equipment, port installations, and maintenance stations.

Critical Infrastructure

The term *critical infrastructure* has been widely adopted to distinguish those infrastructure elements that, if significantly damaged or destroyed, would cause serious disruption of the dependent system or organisation. Storm, flood, or earthquake damage leading to loss of certain transportation routes in a city, for example bridges crossing a river, could make it impossible for people to evacuate, and for emergency services to operate; these routes would be deemed critical infrastructure. Similarly, an on-line booking system might be critical infrastructure for an airline.

Urban Infrastructure

Urban or *municipal infrastructure* refers to hard infrastructure systems generally owned and operated by municipalities, such as streets, water distribution, and sewers. It may also include some of the facilities associated with soft infrastructure, such as parks, public pools and libraries.

Green Infrastructure

Green infrastructure is a concept that highlights the importance of the natural environment in decisions about land use planning. In particular there is an emphasis on the "life support" functions provided by a network of natural ecosystems, with an emphasis on interconnectivity to support long-term sustainability. Examples include clean water and healthy soils, as well as the more anthropocentric functions such as recreation and providing shade and shelter in and around towns and cities. The concept can be extended to apply to the management of stormwater runoff at the local level through the use of natural systems, or engineered systems that mimic natural systems, to treat polluted runoff.

Marxism

In Marxism, the term infrastructure is sometimes used as a synonym for "base" in the dialectic synthetic pair *base and superstructure*. However the Marxist notion of base is broader than the non-Marxist use of the term infrastructure, and some soft infrastructure, such as laws, governance, regulations and standards, would be considered by Marxists to be part of the superstructure, not the base.

Other Uses

In other applications, the term infrastructure may refer to information technology, informal and formal channels of communication, software development tools, political and social networks, or beliefs held by members of particular groups. Still underlying these more conceptual uses is the idea that infrastructure provides organising structure and support for the system or organisation it serves, whether it is a city, a nation, a corporation, or a collection of people with common interests. Examples include IT infrastructure, research infrastructure, terrorist infrastructure, and tourism infrastructure.

Related Concepts

The term *infrastructure* is often confused with the following overlapping or related concepts.

Land Improvement and Land Development

The terms *land improvement* and *land development* are general terms that in some contexts may include infrastructure, but in the context of a discussion of infrastructure would refer only to smaller scale systems or works that are not included in infrastructure because they are typically limited to a single parcel of land, and are owned and operated by the land owner.

For example, an irrigation canal that serves a region or district would be included with infrastructure, but the private irrigation systems on individual land parcels would be considered land improvements, not infrastructure. Service connections to municipal service and public utility networks would also be considered land improvements, not infrastructure.

Public Works and Public Services

The term *public works* includes government owned and operated infrastructure as well as public buildings such as schools and court houses. Public works generally refers to physical assets needed to deliver *public services*. Public services include both infrastructure and services generally provided by government.

Typical Attributes

Hard infrastructure generally has the following attributes.

Capital Assets that Provide Services

These are physical assets that provide services. The people employed in the hard infrastructure sector generally maintain, monitor, and operate the assets, but do not offer services to the clients or users of the infrastructure. Interactions between workers and clients are generally limited to administrative tasks concerning ordering, scheduling, or billing of services.

Large Networks

These are large networks constructed over generations, and are not often replaced as a whole system. The network provides services to a geographically defined area, and has a long life because its service capacity is maintained by continual refurbishment or replacement of components as they wear out.

Historicity and Interdependence

The system or network tends to evolve over time as it is continuously modified, improved, enlarged, and as various components are rebuilt, decommissioned or adapted to other uses. The system components are interdependent and not usually capable of subdivision or separate disposal, and consequently are not readily disposable within the commercial marketplace. The system interdependency may limit a component life to a lesser period than the expected life of the component itself.

Natural Monopoly

The systems tend to be natural monopolies, insofar that economies of scale means that multiple agencies providing a service are less efficient than would be the case if a single agency provided the service. This is because the assets have a high initial cost and a value that is difficult to determine. Once most of the system is built, the marginal cost of servicing additional clients or users tends to be relatively inexpensive, and may be negligible if there is no need to increase the peak capacity or the geographical extent of the network.

In public economics theory, infrastructure assets such as highways and railways tend to be public goods, in that they carry a high degree of non-excludability, where no household can be excluded from using it, and non-rivalry, where no household can reduce another from enjoying it. These properties lead to externality, free ridership, and spillover effects that distort perfect competition and market efficiency. Hence, government becomes the best actor to supply the public goods.

Economics, Management, Engineering, and Impacts

The following concerns mainly hard infrastructure and the specialised facilities used for soft infrastructure.

Ownership and Financing

Infrastructure may be owned and managed by governments or by private companies, such as public utility or railway companies. Generally, most roads, major ports and airports, water distribution systems and sewage networks are publicly owned, whereas most energy and telecommunications networks are privately owned. Publicly owned infrastructure may be paid for from taxes, tolls, or metred user fees, whereas private infrastructure is generally paid for by metred user fees. Major investment projects are generally financed by the issuance of long-term bonds.

An interesting comparison between privatisation versus government-sponsored public works involves high speed rail (HSR) projects in East Asia. In 1998, the Taiwan government awarded the Taiwan High Speed Rail Corporation, a private organisation, to construct the 345 km line from Taipei to Kaohsiung in a 35-year concession contract. Conversely, in 2004 the South Korean government charged the Korean High Speed Rail Construction Authority, a public entity, to construct its high speed rail line, 412 km from Seoul to Busan, in two phases. While different implementation strategies,

Taiwan successfully delivered the HSR project in terms of project management (time, cost, and quality), whereas South Korea successfully delivered its HSR project in terms of product success (meeting owners' and users' needs, particularly in ridership). Additionally, South Korea successfully created a technology transfer of high speed rail technology from French engineers, essentially creating an industry of HSR manufacturing capable of exporting knowledge, equipment, and parts worldwide.

Henceforth, government owned and operated infrastructure may be developed and operated in the private sector or in public-private partnerships, in addition to in the public sector. In the United States, public spending on infrastructure has varied between 2.3% and 3.6% of GDP since 1950. Many financial institutions invest in infrastructure.

Role of Pension Funds

The average allocation to infrastructure only represents 1% of total assets under management by pensions- excluding indirect investment through ownership of stocks of listed utility and infrastructure companies. But there are wide typology differences across regions with many large, sophisticated, pension funds in jurisdictions such as Ontario, Quebec, California, Holland, and Australia already investing more than 5% of their total assets (and typically more than a third of their "alternative" assets) in infrastructure. In countries such as the US, Mexico, Sweden and Norway, there has been a rapid rise of the allocation to infrastructure since 2010, even among more traditional pension funds.

Most pension funds have long-dated liabilities, with matching long-term investments. These large institutional investors need to protect the long-term value of their investments from inflationary debasement of currency and market fluctuations, and provide recurrent cash flows to pay for retiree benefits in the short-medium term: from that perspective, infrastructure is an ideal asset class that provides tangible advantages such as long duration (thus facilitating cash flow matching with long-term liabilities), protection against inflation and statistical diversification (low correlation with 'traditional' listed assets such as equity and fixed income investments), thus reducing overall portfolio volatility.

The various types of pension plans (public and private pensions and superannuation schemes) account for approximately 40% of all investors in the infrastructure asset class, excluding projects directly funded and developed by governments and public authorities.

Infrastructure Asset Management

The method of *infrastructure asset management* is based upon the definition of a Standard of service (SoS) that describes how an asset will perform in objective and measurable terms. The SoS includes the definition of a minimum condition grade, which is established by considering the consequences of a failure of the infrastructure asset.

The key components of infrastructure asset management are:

- Definition of a standard of service
- Establishment of measurable specifications of how the asset should perform
- Establishment of a minimum condition grade
- Establishment of a whole-life cost approach to managing the asset
- Elaboration of an *Asset Management Plan*

The 2009 report card produced by the American Society of Civil Engineers gave America's Infrastructure a grade of "D".

Engineering

Most infrastructure is designed by engineers, urbanists or architects. Generally road and rail transport networks, as well as water and waste management infrastructure are designed by civil engineers, electrical power and lighting networks are designed by power engineers and electrical engineers, and telecommunications, computing and monitoring networks are designed by systems engineers.

In the case of urban infrastructure, the general layout of roads, sidewalks and public places may sometimes be designed by urbanists or architects, although the detailed design will still be performed by civil engineers. If a building is required, it is designed by an architect, and if an industrial or processing plant is required, it may be designed by industrial engineer or a process engineer.

Figure: *The Berlin Brandenburg Airport under construction.*

In terms of engineering tasks, the design and construction management process usually follows these steps:

Preliminary Studies

- Determine existing and future traffic loads, determine existing capacity, and estimate the existing and future standards of service
- Conduct a preliminary survey and obtain information from existing air photos
- Identify possible conflicts with other assets or topographical features
- Perform environmental impact studies:
- Evaluate the impact on the human environment (noise pollution, odors, electromagnetic interference, etc)
- Evaluate the impact on the natural environment (disturbance of natural ecosystems)
- Evaluate the possible presence of contaminated soils;
- Given various time horizons, standards of service, environmental impacts, and conflicts with existing structures or terrain, propose various preliminary designs
- Estimate the costs of the various designs, and make recommendations.

Detailed Survey

- Perform a detailed survey of the construction site
- Obtain "as built" drawings of existing infrastructure
- Dig exploratory pits where required to survey underground infrastructure
- Perform a geotechnical survey to determine the bearing capacity of soils and rock
- Perform soil sampling and testing to estimate nature, degree and extent of soil contamination.

Detailed Engineering

- Prepare detailed plans and technical specifications
- Prepare a detailed bill of materials
- Prepare a detailed cost estimate
- Establish a general work schedule.

Authorisation

- Obtain authorisation from environmental and other regulatory agencies

- Obtain authorisation from any owners or operators of assets affected by the work
- Inform emergency services, and prepare contingency plans in case of emergencies.

Tendering

- Prepare administrative clauses and other tendering documents
- Organise and announce a call for tenders
- Answer contractor questions and issue addenda during the tendering process
- Receive and analyse tenders, and make a recommendation to the owner.

Construction Supervision

- Once the construction contract has been signed between the owner and the general contractor, all authorisations have been obtained, and all pre-construction submittals have been received from the general contractor, the construction supervisor issues an "Order to begin construction"
- Regularly schedule meetings and obtain contact information for the general contractor (GC) and all interested parties
- Obtain a detailed work schedule and list of subcontractors from the GC
- Obtain detailed traffic diversion and emergency plans from the GC
- Obtain proof of certification, insurance and bonds
- Examine shop drawings submitted by the GC
- Receive reports from the materials quality control lab
- When required, review Change requests from the GC, and issue construction directives and change orders
- Follow work progress and authorise partial payments
- When substantially completed, inspect the work and prepare a list of deficiencies
- Supervise testing and commissioning
- Verify that all operating and maintenance manuals, as well as warranties, are complete
- Prepare "as built" drawings
- Make a final inspection, issue a certificate of final completion, and authorise the final payment.

Impact on Economic Development

Investment in infrastructure is part of the capital accumulation required for economic development and may have an impact on socioeconomic measures of welfare. The causality of infrastructure and economic growth has always been in debate. In developing nations, expansions in electric grids, roadways, and railways show marked growth in economic development. However, the relationship does not remain in advanced nations who witness more and more lower rates of return on such infrastructure investments.

Nevertheless, infrastructure yields indirect benefits through the supply chain, land values, small business growth, consumer sales, and social benefits of community development and access to opportunity. The American Society of Civil Engineers cite the many transformative projects that have shaped the growth of the United States including the Transcontinental Railroad that connected major cities from the Atlantic to Pacific coast; the Panama Canal that revolutionized shipment in connected the two oceans in the Western hemisphere; the Interstate Highway System that spawned the mobility of the masses; and still others that include the Hoover Dam, Trans-Alaskan pipeline, and many bridges (the Golden Gate, Brooklyn, and Bay Bridge). All these efforts are testimony to the infrastructure and economic development correlation.

Use as Economic Stimulus

During the Great Depression of the 1930s, many governments undertook public works projects in order to create jobs and stimulate the economy. The economist John Maynard Keynes provided a theoretical justification for this policy in *The General Theory of Employment, Interest and Money*, published in 1936. Following the global financial crisis of 2008–2009, some again proposed investing in infrastructure as a means of stimulating the economy.

Environmental Impacts

While infrastructure development may initially be damaging to the natural environment, justifying the need to assess environmental impacts, it may contribute in mitigating the "perfect storm" of environmental and energy sustainability, particularly in the role transportation plays in modern society. Offshore wind power in England and Denmark may cause issues to local ecosystems but are incubators to clean energy technology for the surrounding regions. Ethanol production may overuse available farmland in Brazil but have propelled the country to energy independence. High speed rail may cause noise and wide swathes of rights-of-way through countrysides and urban communities but have helped China, Spain, France, Germany, Japan, and other nations deal with concurrent issues of economic competitiveness, climate change, energy use, and built environment sustainability.

History

Before 1700: Infrastructure before 1700 consisted mainly of roads and canals. Canals were used for transportation or for irrigation. Sea navigation was aided by ports and lighthouses. A few advanced cities had aqueducts that serviced public fountains and baths, while fewer had sewers.

Roads: The first roads were tracks that often followed game trails, such as the Natchez Trace.

The first paved streets appear to have been built in Ur in 4000 BCE. Corduroy roads were built in Glastonbury, England in 3300 BCE and brick-paved roads were built in the Indus Valley Civilization on the Indian subcontinent from around the same time. In 500 BCE, Darius I the Great started an extensive road system in Persia (Iran), including the Royal Road. With the rise of the Roman Empire, the Romans built roads using deep roadbeds of crushed stone as an underlying layer to ensure that they kept dry. On the more heavily travelled routes, there were additional layers that included six sided capstones, or pavers, that reduced the dust and reduced the drag from wheels.

In the medieval Islamic world, many roads were built throughout the Arab Empire. The most sophisticated roads were those of the Baghdad, Iraq, which were paved with tar in the 8th century.

Canals and Irrigation Systems

The oldest known canals were built in Mesopotamia circa 4000 BCE, in what is now modern day Iraq and Syria. The Indus Valley Civilization in India and Pakistan from c3300 BCE had a sophisticated canal irrigation system. In Egypt, canals date back to at least 2300 BCE, when a canal was built to bypass the cataract on the Nile near Aswan. In ancient China, large canals for river transport were established as far back as the Warring States (481-221 BCE). By far the longest canal was the Grand Canal of China completed in 609 CE, still the longest canal in the world today at 1,794 kilometres (1,115 mi).

In Europe, canal building began in the Middle Ages because of commercial expansion from the 12th century CE. Notable canals were the Stecknitz Canal in Germany in 1398, the Briare Canal connecting the Loire and Seine in Francein 1642, followed by the Canal du Midi in 1683 connecting the Atlantic to the Mediterranean. Canal building progressed steadily in Germany in the 17th and 18th centuries with three great rivers, the Elbe, Oder, and Weser being linked by canals.

1700 to 1870

Roads: As traffic levels increased in England and roads deteriorated, toll roads were built by *Turnpike Trusts*, especially between 1730–1770. Turnpikes were also later built in the United States. They were usually built by private companies under a government franchise. Water transport on rivers and canals carried many farm goods from the US frontier between the Appalachian Mountains and Mississippi River in the early 19th century, but the shorter road route over the mountains had advantages.

In France, Pierre-Marie-Jérôme Trésaguet is widely credited with establishing the first scientific approach to road building about the year 1764. It involved a layer of large rocks, covered by a layer of smaller gravel. John Loudon McAdam (1756–1836) designed the first modern highways, and developed an inexpensive paving material of soil and stone aggregate known as macadam.

Canals: In Europe, particularly Britain and Ireland, and then in the early US and the Canadian colonies, inland canals preceded the development of railroads during the earliest phase of the Industrial Revolution. In Britain between 1760 and 1820 over one hundred canals were built.

In the United States, navigable canals reached into isolated areas and brought them in touch with the world beyond. By 1825 the Erie Canal, 363 miles (584 km) long with 82 locks, opened up a connection from the populated northeast to the fertile Great Plains. During the 19th century, the length of canals grew from 100 miles (160 km) to over 4,000 miles (6,400 km), with a complex network in conjunction with Canada making the Great Lakes navigable, although some canals were later drained and used as railroad rights-of-way.

Railways: The earliest railways were used in mines or to bypass waterfalls, and were pulled by horses or by people. In 1811 John Blenkinsop designed the first successful and practical railway locomotive, and a line was built connecting the Middleton Colliery to Leeds. The Liverpool and Manchester Railway, considered to be the world's first intercity line, opened in 1826. In the following years, railways spread throughout the United Kingdom and the world, and became the dominant means of land transport for nearly a century.

In the US, the 1826 Granite Railway in Massachusetts was the first commercial railroad to evolve through continuous operations into a common carrier. The Baltimore and Ohio, opened in 1830, was the first to evolve into a major system. In 1869, the symbolically important transcontinental railroad was completed in the US with the driving of a golden spike at Promontory, Utah.

Telegraph Service: The electrical telegraph was first successfully demonstrated on 25 July 1837 between Euston and Camden Town in London. It entered commercial use on the Great Western Railway over the 13 miles (21 km) from Paddington station to West Drayton on 9 April 1839.

In the United States, the telegraph was developed by Samuel Morse and Alfred Vail. On 24 May 1844, Morse made the first public demonstration of his telegraph by sending a message from the Supreme Court Chamber in the US Capitol in Washington, DC to the B&O Railroad outer depot (now the B&O Railroad Museum) in Baltimore. The Morse/Vail telegraph was quickly deployed in the following two decades. On 24 October 1861, the first transcontinental telegraph system was established. The first successful transatlantic telegraph cable was completed on 27 July 1866, allowing transatlantic telegraph communications for the first time. Within 29 years of its first installation at Euston Station, the telegraph network crossed the oceans to every continent but Antarctica, making instant global communication possible for the first time.

1870 to 1920

Roads: Tar-bound macadam, or tarmac, was applied to macadam roads towards the end of the 19th century in cities such as Paris. In the early 20th century tarmac and concrete paving were extended into the countryside.

Canals: Many notable sea canals were completed in this period, such as the Suez Canal in 1869, the Kiel Canal in 1897, and the Panama Canal in 1914.

Telephone Service: In 1876, Alexander Graham Bell achieved the first successful telephone transmission of clear speech. The first telephones had no network, but were in private use, wired together in pairs. Users who wanted to talk to different people had as many telephones as necessary for the purpose. A user who wished to speak, whistled into the transmitter until the other party heard. Soon, however, a bell was added for signalling, and then a switch-hook, and telephones took advantage of the exchange principle already employed in telegraph networks. Each telephone was wired to a local telephone exchange, and the exchanges were wired together with trunks. Networks were connected together in a hierarchical manner until they spanned cities, countries, continents, and oceans.

Electricity: At the Paris Exposition of 1878, electric arc lighting had been installed along the Avenue de l'Opera and the Place de l'Opera, using electric Yablochkov arc lamps, powered by Zénobe Gramme alternating current dynamos.

Yablochkov candles required high voltages, and it was not long before experimenters reported that the arc lights could be powered on a seven mile (11 km) circuit. Within a decade scores of cities would have lighting systems using a central power plant that provided electricity to multiple customers via electrical transmission lines. These systems were in direct competition with the dominant gaslight utilities of the period. The first electricity system supplying incandescent lights was built by the Edison Illuminating Company in lower Manhattan, eventually serving one square mile with six "jumbo dynamos" housed at Pearl Street Station.

The first transmission of three-phase alternating current using high voltage took place in 1891 during the International Electro-Technical Exhibition in Frankfurt. A 25 kilovolt transmission line, approximately 175 km (109 mi) long, connected Lauffen on the Neckar with Frankfurt. Voltages used for electric power transmission increased throughout the 20th century. By 1914 fifty-five transmission systems operating at more than 70,000 V were in service, the highest voltage then being used was 150,000 V.

Water Distribution and Sewers: In the 19th century major treatment works were built in London in response to cholera threats. The *Metropolis Water Act (1852)* was enacted. "Under the Act, it became unlawful for any water company to extract water for domestic use from the tidal reaches of the Thames after 31 August 1855, and from 31 December 1855 all such water was required to be effectively filtered. The *Metropolitan Commission of Sewers* was formed, water filtration was made compulsory, and new water intakes on the Thames were established above Teddington Lock. The technique of purification of drinking water by use of compressed liquefied chlorine gas was developed in 1910 by US Army Major Carl Rogers Darnall, Professor of Chemistry at the Army Medical School. Darnall's work became the basis for present day systems of municipal water purification.

Subways: In 1863 the London Underground was created. In 1890, it first started using electric traction and deep-level tunnels. Soon afterwards, Budapest and many other cities started using subway systems. By 1940, nineteen subway systems were in use.

Since 1920

Roads: In 1925, Italy was the first country to build a freeway-like road, which linked Milan to Lake Como, known as the Autostrada dei Laghi. In Germany, the autobahns formed the first limited-access, high-speed road network in the world, with the first section from Frankfurt am Main to Darmstadt opening in 1935. The first long-distance rural freeway in the United States is generally considered to be the Pennsylvania Turnpike, which opened on October 1, 1940. In the United States, the Interstate Highway System was authorised by the Federal-Aid Highway Act of 1956. Most of the system was completed between 1960 and 1990.

Infrastructure in the Developing World: According to researchers at the Overseas Development Institute, the lack of infrastructure in many developing countries represents one of the most significant limitations to economic growth and achievement of the Millennium Development Goals (MDGs). Infrastructure investments and maintenance can be very expensive, especially in such as areas as landlocked, rural and sparsely populated countries in Africa.

It has been argued that infrastructure investments contributed to more than half of Africa's improved growth performance between 1990 and 2005, and increased investment is necessary to maintain growth and tackle poverty.

The returns to investment in infrastructure are very significant, with on average thirty to forty percent returns for telecommunications (ICT) investments, over forty percent for electricity generation, and eighty percent for roads.

Regional Differences: The demand for infrastructure, both by consumers and by companies is much higher than the amount invested. There are severe constraints on the supply side of the provision of infrastructure in Asia. The infrastructure financing gap between what is invested in Asia-Pacific (around US$48 billion) and what is needed (US$228 billion) is around US$180 billion every year.

In Latin America, three percent of GDP (around US$71 billion) would need to be invested in infrastructure in order to satisfy demand, yet in 2005, for example, only around two percent was invested leaving a financing gap of approximately US$24 billion.

In Africa, in order to reach the seven percent annual growth calculated to be required to meet the MDGs by 2015 would require infrastructure investments of about fifteen percent of GDP, or around US$93 billion a year. In fragile states, over thirty-seven percent of GDP would be required.

Sources of Funding: Currently, the source of financing varies significantly across sectors. Some sectors are dominated by government spending, others by overseas development aid (ODA), and yet others by private investors.

In sub-Saharan Africa, the government spends around US$9.4 billion out of a total of US$24.9 billion. In irrigation, governments represent almost all spending. In transport and energy a majority of investment is government spending. In ICT and water supply and sanitation, the private sector represents the majority of capital expenditure. Overall, between them aid, the private sector, and non-OECD financiers exceed government spending. The private sector spending alone equals state capital expenditure, though the majority is focused on ICT infrastructure investments. External financing increased in the 2000s and in Africa alone external infrastructure investments increased from US$7 billion in 2002 to US$27 billion in 2009. China, in particular, has emerged as an important investor.

Bibliography

Asaithambi S. : *Economics of Ground Water Management in India*, Abhijeet, Delhi, 2008.

Bhave, P.R. and R. Gupta: *Analysis of Water Distribution Networks*, Narosa, Delhi, 2011.

Biswas Asit K. : *Integrated Water Resources Management in South and South-East Asia*, , Oxford University Press, Delhi, 2001.

Boyer , J.S.: *Measuring the Water Status of Plants and Soils,* Academic Press, N.Y., 1995.

Conway, D.: *Climate Change and Water Resources in the Nile Basin*, University of London, London, 1993.

Cronon, William: *Changes in the Land: Indians, Colonists,. and the Ecology of New England.* New York: Hill and Wang, 1983.

Crouch, D. P.: *Water Management in Ancient Greek Cities.* New York: Oxford University Press, 1993.

Darwish, Adel: *Water Wars: Coming Conflicts in the Middle East*, London, Victor Gollancz, 1999.

Doneen L.D.: *Water Quality for Agriculture*, Department of Irrigation, University of California, Davis, 1964.

Easter, K. William : : *Irrigation Investment Technology, and Management Strategies for Development Studies in Water Policy Management 9*, Boulder, Colo Westview Press, 1986.

Gaur, P.K. : *Principles of Water and Wastewater Treatment*, SBS Publishers, Delhi, 2012.

Hammer, Mark J.: *Water& Wastewater Technology*. John Wiley& Sons, Inc., New York, 1975.

Hillel, Daniel: *Rivers of Eden: the Struggle for Water and the Quest for Peace in the Middle East*, Oxford University Press, NYC, 1994.

Husain, Ahmad : *Environment and Water Resource Management*, Sumit Enterprises, Delhi, 2006.

Issar, Arie: *Water Shall Flow from the Rock: Hydrogeology and Climate in the Lands of the Bible*, Springer-Verlag, NYC, 1990.

Jana B L : *Water Harvesting and Watershed Management*, Agrotech, Delhi, 2008.

Jat M.L. and Bhakar S.R. : *Ground Water Hydrology : Theory and Practice*, Agrotech, Delhi, 2009.

Katz, J.: *Ozone and Chlorine Dioxide Technology of Disinfection of Drinking Water,* Noyes Data Corp., Park Ridge, New Jersey, 1980.

Kaul, S.N. : *Technological Applications in Wastewater Engineering*, Daya, Delhi, 2009.

Maier, F. J.: *Manual of Water Fluoridation Practice*. McGraw-Hill Book Co,, New York, 1963.

Matsow R.: *Water in the Middle East: Conflict or Cooperation*, Boulder, 1990.

Meenu Bhatnagar: *Groundwater Management : Sustainable Approaches*, Icfai Books, Delhi, 2012.

Merrett, Stephen, *Water for Agriculture: Irrigation Economics in International Perspective*, Spon Press, London, 2001.

Michael John: *Earth's Changing Surface: an Introduction to Geomorphology*, Clarendon Press, Oxford, 1985.

Nemerow, N. L.: *Liquid Wastes of Industry*; Addison-Wesley Publishing Co. Reading, Mass; U.S.A., 1971.

Okun, Daniel A.: *Elements of Water Supply and Wastewater Disposal,* John Wiley & Sons, New York, 1971.

Pandey, Mahendra: *Wastewater Management*, Dominant, Delhi, 2011.

Paul A. Keddy: *Wetland Ecology – Principles and Conservation*, Cambridge University Press, Cambridge, 2000.

Postel, Sandra : *Last Oasis: Facing Water Scarcity*, World Water Institute, New York, 1992.

Robbins, F. W.: *The Story of Water Supply*. London, U.K.: Oxford University Press, 1946.

Shamsi, U.M. : *GIS Applications for Water, Wastewater, and Stormwater Systems*, Routledge, Delhi, 2010.

Siddharth Kaul: *Low Cost Wastewater Treatment Technologies*, ABD Publishers, Delhi, 2001.

Spiegler, K. S.: *Principles of Desalination*. Academic Press, New York, 1966.

Tideman, E. M.: *Watershed Management – Guidelines for Indian Conditions*, Omega Scientific Publishers, Delhi, 1996.

Whipple, G. C., Fair G. M., and Whipple, M. C.: *Microscopy of Drinking Water,* John Wiley & Sons, New York, 1947.

Wolfson, Lois G.: *Rural Groundwater Contamination*. Lewis Publishers, Chelsea, 1987.

Index

❑❑❑